Leadership With A Purpose:

Motivating Your Engineers

Robert D. Murphy

CRYPTO FLIGHT LLC

Florida

Copyright 2022 © Robert D. Murphy

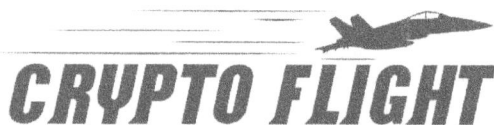

CRYPTO FLIGHT

Published by Crypto Flight LLC

For more information, contact: hello@cryptoflight.io

ISBN (electronic book):   979-8-9874420-0-5
ISBN (print):   979-8-9874420-1-2

Printed in the U.S.A.

TO MY BEST FRIEND, MY MOTHER, LOIS.

The game of life could not have been won

Without the woman from Deland who loves her son.

Love Robbie

Table of Contents

Introduction

In today's world, effectively motivating and managing highly intelligent technology professionals is challenging. Around the world, people are more educated, informed and highly social about their careers and employers, making retention and employee happiness a full-time job. In fact, companies are beginning to recognize these impacts and have begun employing more "PeopleOps" type human resources professionals to tackle this head-on.

The reality is that most engineers don't leave jobs, they leave managers. While having a swanky and modern human resources strategy helps keep the company on the cutting edge of benefits and overall happiness, it does not solve the root problem – having good managers that motivate engineering teams.

In the remote and global workforce that we all participate in, establishing a healthy work culture with safe boundaries is an ongoing task that cannot be forced. Add in the complexities of global cultural differences and political phenomena; it's every leader's nightmare to manage well.

This book will cover what makes an engineering culture, how to cultivate it, and how to build trust with your teams.

After reading this book, you can effectively motivate, inspire and lead engineering professionals of all levels that are happy and loyal to you.

About The Author

When asked about the greatest motivation in his life, Robert Murphy reflected upon his career and dedicated that title to his mother. At the young age of 14, he stepped into the world of software engineering as an amateur programmer. Robert's mother wholeheartedly supported his ventures, and he eventually became an entrepreneur of online text-based games. It was the combination of early-on successes and his mother's undying encouragement that fueled his ambition and passion for the industry.

Backed by years of experience within the technology sector, Robert has served in various capacities that added to his prowess in software engineering, business development, team building, and people management. He has built his own successful businesses from the ground up and played a pivotal part in the growth of other companies. Throughout multiple leadership roles, Robert noticed one secret ingredient that made all the difference in the propensity of any business— company culture.

Celebrated for his people-first leadership style, Robert was often referred to as 'The King of Culture' by a previous employer. From proper recognition and compensation to expert-level professional development, he became renowned for his ability to develop staff talent and leverage that expertise for the betterment of any company. Robert once experienced the strife many inexperienced leaders face in the software engineering field, including apathetic and unmotivated team members.

With a passion for helping others, Robert has now turned his attention to helping leaders within the technology field establish healthy company cultures and motivate intelligent but underperforming, engineering professionals. Although *Leadership with a Purpose: Motivating Your Engineers* is his first book, he has written numerous blog posts and case studies on the subject.

When he is away from the office, Robert enjoys relaxing by heading outdoors to fish or hunt. As a pilot, he also loves dedicating much of his time to maintaining his skills as an aviator. He can often be found heading out on flights to explore the country. When the weather isn't cooperating, he feels content to stay at home in his house located in Central Florida to experience lazy mornings, evenings in the hot tub, and a single malt scotch by the fire. Even when he is not working, Robert is consistently researching his industry to stay updated on today's best practices in engineering and proven trends in culture building.

Chapter 1

Introducing Engineering

Culture

Engineering culture is a complex phenomenon that's created by engineers, for engineers. The culture sets the tone and behavior of an organization from the top to the support staff. It includes how we interact with one another on a day-to-day basis, our interactions with customers and suppliers, and how we handle mistakes and problems.

You may think that engineers want to "get the job done," but good engineering cultures go beyond that. They motivate engineers to be creative and innovative, encouraging them to take ownership of their work and find efficient and effective solutions. Good engineering cultures also create an environment where team members can learn from each other in a safe and supportive way. I like engineering cultures as an ecosystem where everyone is rewarded for their contributions and ideas. You know you have a good engineering culture when everyone pulls in the same direction and works towards common goals.

The "Culture" Buzzword

We often hear the "culture" buzzword when discussing engineering cultures. It can be confusing because culture is a broad and complex concept that's difficult to define. The simplest way to think about it is as an organization or group of people's collective beliefs, values, practices, and behaviors. Culture brings people together and sets expectations for their behavior in different situations.

In the business world, culture is often used to describe a company's values and goals that shape its decision-making process. It can also be used to measure employee satisfaction and engagement or to understand why specific strategies are more successful than others.

At its core, engineering culture is an extension of company culture. It defines how engineers think and interact with each other and sets expectations for how they should behave when working together. When well-defined, it encourages innovation and creativity while maintaining high standards of quality and professionalism. Have you ever been part of an engineering team with a robust culture? If so, then you know what I mean!

The way you cultivate and manage engineering culture has a significant impact on your team's performance. I like to think of the process like this: Engineering culture is like rocket fuel. You can use it to help a team push themselves to do amazing things, or you can overdo it and cause some severe injuries.

Great engineering cultures are built from the ground up, starting with your people. To understand how engineers think, what drives their behavior, and how they interact with other engineers within your company, you must first delve into engineering culture before you even get started on management techniques.

Still on the culture buzzword, let's discuss some key aspects of engineering culture

Workplace culture is the values and behaviors that characterize a company or organization. It includes its mission, beliefs, policies, practices, and employee expectations.

Workplace culture is either positive or negative, depending on how management develops and supports it. When a workplace culture is positive, it can foster strong relationships between coworkers and increase job satisfaction, loyalty, and creativity. When negative, it can create an uncomfortable work environment that is not conducive to productivity or team morale. Its discomfort is identified simply by observing cynicism. If you have lots of negative or passive-aggressive attitudes that are cynical, it is a sign your culture is working against you.

Unfortunately, some view workplace culture as a half-hearted buzzword used to influence potential and current team members' loyalty to the company rather than something meaningful or valuable. However, having a strong workplace culture is essential for any organization to thrive. It should be created through meaningful and open dialogue with team members, customers, and other stakeholders. When everyone feels respected and valued within the company's environment, productivity and retention will improve for both team members and employers. It ultimately leads to tremendous success for everyone involved.

Team culture is the values and behaviors shared among members of a specific group or team. These cultures are usually more personal than workplace cultures because they rely on individuals' feelings, beliefs, and experiences to create cohesion. Team cultures can be positive or negative depending on the dynamics within the team. When a team is supportive and trusting, it can foster creativity, productivity, and innovation. On the other hand, when team dynamics are strained or hostile, it can cause friction and negatively affect performance.

Creating a solid team culture starts with understanding the individual members of the team and their unique perspectives. It means that each person should feel comfortable speaking up about their opinions and ideas without fear of judgment. Additionally, team members should be encouraged to collaborate and work together towards a common goal. Finally, teams should have clearly defined roles and responsibilities so that everyone understands their part in the team's success. If you've found a team with this culture, hang on to it. It's precious!

When people hear the buzzword "workplace culture," they may feel excitement because they think it will be a good working environment with positive relationships. However, if they find out the culture is fake and insincere, it can make them feel used and unvalued. So, companies need to foster and create a genuine workplace culture that is honest and reflects the company's values

Building a solid workplace culture starts with setting clear expectations for team members. Management should communicate their vision, mission, and values meaningfully so everyone understands what is expected from them regarding behavior, attitude, and performance. Furthermore, it is essential to note that these expectations are non-negotiable.

Creating the right environment is also key to a thriving workplace culture. Management should create an atmosphere of collaboration and trust by providing resources, tools, and training. It will help team members feel comfortable expressing themselves and their ideas without fear of judgment or criticism. Creating an atmosphere of openness can cultivate a strong sense of team spirit and camaraderie.

To ensure that the workplace culture is maintained over time, management should regularly assess and evaluate it. It means measuring employee engagement, evaluating customer feedback, and making changes if necessary. It also requires recognizing team members for their excellent work and providing rewards to motivate them. If you invest in a strong workplace culture, your team members will go the extra mile for you!

Having a positive and productive workplace culture is essential for any organization. It can increase employee satisfaction, loyalty, and productivity. Building a stable workplace climate will pay dividends in team morale, customer satisfaction, and overall business performance. So don't underestimate the power of workplace culture! It does make a difference.

When most people think of workplace culture, they think of an artificial or forced culture. It is when management tries to create a specific culture to influence employee behavior and performance. However, another type of workplace culture is organic, natural, and cultivated without outside intervention. It is when the company's culture evolves due to its team members' collective actions and relationships.

A workplace culture that is organic and natural has a strong sense of identity and community. Employees feel part of something larger than themselves and are proud to be associated with the company. This culture is usually more productive and harmonious because team members share common values and beliefs.

On the other hand, a workplace culture cultivated without outside intervention can be harmful if it's not managed correctly. It is when the company's culture becomes twisted, and team members start behaving in ways that are not conducive to productivity or team morale. For example, some companies may have a "no-failure" policy where team members are afraid to take risks for fear of being reprimanded, or team members may compete against each other instead of working together.

Ultimately, it's up to management to foster and create a workplace culture that is positive, productive, and reflects the values of the company. By setting clear expectations, creating a supportive environment, and regularly assessing and evaluating the culture, you can ensure that your company has a thriving workplace culture that benefits everyone involved!

Here is a practical example of what I'm trying to say:

Recently, a company implemented a new strategy to foster team collaboration and creativity. They created an open environment where everyone's ideas were welcomed and encouraged, with rewards given for innovation. It has helped team members feel more comfortable working together and sharing thoughts and opinions without fear of criticism or failure. Moreover, team members now work together to develop creative solutions and are more motivated than ever. The overall atmosphere in the company has improved significantly, resulting in higher productivity and morale.

An organic, natural workplace culture can produce positive results for everyone involved. So, if you want a productive and happy team, take the time to invest in your workplace culture today. You won't regret it!

There are two types of workplace cultures: the kind that's forced and the kind that evolves organically. The forced kind is when management tries to create a specific culture to influence employee behavior and performance. However, the organic kind is when the company's culture evolves as a result of the collective actions and relationships of its team members.

Many companies have phony cultures where team members are afraid to take risks or feel uncomfortable sharing their thoughts and opinions. If you see these signs, it's best not to work there. On the other hand, a company with an organic culture cultivated without outside intervention can be damaging if it's not managed correctly. It is when the company's culture becomes twisted, and team members start behaving in ways that are not conducive to productivity or team morale. When you see these signs, it's usually best to look elsewhere.

I learned an important lesson: You get what you tolerate. If you tolerate a certain level of bad behavior at your job, that's what you'll get. You might be unable to get rid of people who act like jerks or have annoying habits, but you can undoubtedly change the culture in small but significant ways that can have a lasting impact. It is where it becomes vital to have the right mindset.

I can't tell you how often people have asked me, "How do you deal with difficult people?" or "How do you manage a bad culture?" The truth is that there will always be a problem for team members and managers. There will always be reasons why it's impossible to make changes. But if you have the right mindset, you can deal with it.

Here's how I've handled challenging or bad situations in the past:

First, I remind myself that everyone has a role to play on a team. It doesn't mean everyone will agree with my ideas and suggestions. But that's okay because a diversity of thought is essential for us to learn from experience and improve as individuals and team members.

But if someone is always shot-down and their ideas ridiculed to the point of embarrassment, it becomes a problem. Even if you're in a leadership role and everyone else always agrees with you, there's still room for diversity of thought. Everybody is entitled to their opinion and to make suggestions.

Second, I remind myself that the problem isn't the other person or me—our collective inability to work together. I've learned that challenging situations are much easier to deal with when you're not taking things personally. When an employee is arguing with you, it doesn't mean they have something against you—it can just mean they disagree with your position.

Third, I remind myself that a bad workplace culture is garbage, and I will get out of it as soon as possible. If the company isn't doing well, the culture will follow suit. And I want to be with a company that treats its team members well. There are enough companies with toxic cultures that don't put team members first. I will be with a company with a positive culture that treats its people right.

It's important to remember that we're all human beings, and we all have the exact basic needs: time for self-renewal, the opportunity to give and receive love, room for self-expression, growth opportunities, and fair compensation. We want to be with people who make us feel good about ourselves and allow us to grow as individuals.

As you seek a new opportunity, remember that it is okay to say no to a job or company because of its culture. It's okay not to work with people you don't like. Remember that you're building your legacy and creating your legacy—and you get to choose which team you want to belong to. A phony workplace culture will erode your energy and capability for growth, whereas an authentic genuine culture will nourish your and your team's growth.

The importance of a strong engineering culture

Now, let's talk about how you can create a strong engineering culture that motivates. The vital thing to note is that a great engineering culture is not just about having the best tools and methods; it's also about people and relationships.

Engineering culture is essential to any company's success in today's digital world. It is a shared set of engineers' behaviors, attitudes, values, and practices that influence how they think, act and feel. Engineering culture can help businesses become more resilient and reach new heights by empowering technologists to excel and providing guidance for every role.

A strong engineering culture starts with an inclusive workplace where everyone is encouraged to do their best work and push past their boundaries. Senior technologists should provide mentoring opportunities while encouraging organic innovation within the organization. Learning paths are established to provide growth opportunities for all team members while fostering practice communities, making it much easier to retain talent. When people feel a sense of belonging and ownership in their work, they are more likely to be motivated and productive.

Peter Drucker, a renowned management consultant, says, "Culture eats strategy for breakfast." A well-crafted engineering culture is essential to any company's success. Without it, the organization will lack direction and risk stagnating or falling behind the competition. It's also important to constantly reassess your engineering culture to ensure that it meets your technologists' needs.

Having an influential engineering culture goes beyond organizational leaders and branding campaigns, as it must be heard externally by engineers, engineering communities, and social media channels. This way, you can show the world what makes your organization unique and why engineers should want to work with you. Here is an analogy: if the culture is the heart of your organization, then it must be pumping and circulating to bring life to all its parts.

A strong engineering culture is established in an inclusive workplace where:

• Technologists are empowered to achieve and maintain excellence

• Senior technologists provide technical mentoring

• Organic applied innovation from all roles is encouraged

• There are learning paths and communities of practice for every role

• Growth opportunities are available

Let's break each of these down.

Technologists must be given the tools and resources to achieve their goals and stay ahead of the competition. Senior technologists serve as role models who can offer guidance and provide mentorship. It means setting clear expectations, giving constructive feedback, and recognizing accomplishments.

Organic innovation should be encouraged as it fosters creativity and encourages problem-solving. It means providing an environment where everyone can contribute ideas and ask questions without feeling judged or out of place.

Learning paths are integral for team members to acquire new skills and develop professionally. It can include online courses, seminars, workshops, or even hands-on projects that allow technologists to apply their knowledge in the workplace. Additionally, communities of practice offer a great way to share best practices while promoting collaboration across teams.

Finally, growth opportunities are essential for technologists to stay motivated and engaged at work. It could involve salary increases, promotions, or internal and external career development initiatives.

Having an influential engineering culture is essential to any successful organization. It enables technologists to excel, encourages organic innovation, provides employee learning paths, and offers growth opportunities. By fostering a supportive environment that values collaboration and creativity, you can ensure that your organization will remain ahead of the competition in today's ever-evolving digital world.

When developing an engineering culture within your company, it's essential to keep in mind that communication is critical. As mentioned earlier, more is needed to have a strategy - if engineers need to hear the message internally and externally. You will need help to reach its full potential.

Influential engineering culture is vital to attracting and retaining talent. With a well-crafted plan, your organization can create an inclusive workplace where technologists are empowered to succeed. It will help ensure that your organization remains competitive and successful in today's fast-paced market.

Chapter 2

Dealing with "Woke Culture"

Let's talk about the "Woke Culture" phenomenon that has influenced workplace cultures over the past few years. It is defined as the idea that everyone should be aware of and sensitive to social and political issues such as racism, sexual orientation, gender, and identity. In today's workplace, people are more publicly involved in social and political issues, which impacts your team's performance. It is best to have a neutrally grounded stance on handling issues arising from these new social norms.

Getting caught up in the idea that everyone must always agree is easy. That's not necessarily the case. A lot of times, when there is disagreement, it leads to innovative solutions. Instead of trying to silence voices, we should encourage conversations around different points of view. At the same time, not allowing these differences to distract from the business objectives brings everyone together in the first place.

So while it's important to recognize the importance of "woke culture" in the workplace, I think it's also important to remember that not everyone will agree 100% on every issue. Instead of trying to silence voices, we should create an open environment where people can express their views openly and respectfully. It is the only way that everyone can truly benefit from each other's ideas and work together towards solutions that work for all.

How does woke culture cause issues?

While woke culture can be seen as a positive thing, it can also cause issues in the workplace. For one, it can be difficult for employers to balance allowing team members to express themselves and maintaining a professional environment. Additionally, some people may feel uncomfortable or attacked when coworkers start discussing social justice issues.

Ultimately, woke culture in the workplace needs to be handled carefully. Employers should create an environment where team members feel comfortable expressing themselves and ensure that these conversations stay respectful and productive. **Let's look at some strategies for dealing with woke culture in the workplace.**

Set up guidelines

One way to handle woke culture is to set up guidelines. Guidelines are put in place that dictates what topics are allowed to be discussed and which should be avoided. Additionally, employers can have a clear policy on handling disagreements properly. These guidelines will help ensure that conversations remain productive and respectful. For instance, I used to work in a company with a "no-shaming" policy—we agreed not to shame each other for our differing opinions. It made it much easier to have productive conversations.

Encourage open communication

Another way to deal with a woke culture is to encourage open communication. It will help people feel comfortable sharing their views while limiting unproductive conversations and creating a healthy environment for all team members. The easiest way to encourage open communication is through a basic "code of conduct." It should outline the expected level of communication and ensure that everyone is held accountable for their actions.

The code of conduct is used to communicate expectations for different situations, such as how to handle disagreements with coworkers. When creating a code of conduct, employers should note what kinds of conversations are expected and what is not. Let's use the "no shaming" policy as an example. When creating a code of conduct, employers can specify that shaming is not allowed and explain how to handle disagreements to avoid offending people.

Remember, it is essential to put in place a culture where people feel safe expressing their views and disagreeing with others. Still, creating a safe environment for people who may find these conversations uncomfortable is also essential. That's why ensuring everyone understands the guidelines and expectations when working in a company is essential.

Make it clear that everyone has different opinions

I've noticed with the woke culture that some people must agree on social issues to fit in. It isn't necessarily true—everyone has different opinions, and that's okay. Employers should clarify that everyone is allowed to have their own beliefs as long as they are expressed respectfully. It will help create an environment where people can express their views without feeling intimidated or judged.

Foster an environment of respect

You'll agree with me that it's crucial to foster an environment of respect. It may mean having regular meetings to discuss issues in the workplace—this will allow people to express their views without feeling like they are being attacked or judged. Employers should have a clear policy on handling disagreements and remind everyone that attacking another person's opinion is never acceptable. To do this, let's look at how we can use different techniques, such as role-play or active listening:

- **Role-play**: People are asked to play different roles and work through a situation together. It will help them respectfully practice handling disagreements while understanding each other's perspectives.

- **Active listening**: Encourage people to listen attentively before responding, which will help ensure everyone is heard and respected.

- **Giving feedback**: Provide constructive feedback to people and let them know when their behavior is inappropriate.

By creating a culture of respect, employers can ensure that conversations remain productive and respectful, which will ultimately help create an environment where everyone feels comfortable expressing their views.

Hold everyone accountable

Now, let's turn to hold everyone accountable. Employers must make the expectations clear and hold people accountable for their actions. It may mean having a disciplinary policy outlining consequences for inappropriate behavior. Employers should also encourage team members to report any behavior that violates the established code of conduct. It will help create an environment where no one can get away with inappropriate behavior.

Employers can ensure that everyone is heard and respected by creating a culture of respect. It will create an environment where people feel comfortable expressing their views and having meaningful conversations. Ultimately, this will lead to happier team members and more productive work environments!

Model the behavior

Employers should be sure to model the behavior they expect from others. It means leading by example and setting a good example for your team members. Showing respect and empathy towards your colleagues are essential to creating an inclusive culture where everyone feels safe and respected. For instance, employers should be sure to take the time to listen to team members, show empathy when responding, and avoid making assumptions about other people's opinions. Modeling this behavior will help create a culture where everyone is respected and valued.

Creating an engineering culture that motivates is no easy feat, but it's worth the investment. You could be the one to bring out the best in your teams and help them reach their full potential. With these tips, you'll be able to foster an environment of respect, hold everyone accountable, and put in place the behavior you want to see from others. Now that's leadership with a purpose!

The Impact of the Woke Culture on Business Operations When Employees Discuss Social Issues

Still on woke culture, the awakened culture's impact on business operations when team members discuss social issues is undeniable. Woke culture has become popular in the last few years, with many companies now actively encouraging their team members to engage in conversations about social justice and other essential topics.

While these discussions can be beneficial for businesses – allowing them to create a more inclusive work culture and foster open dialogue – they can also cause disruptions in the workplace. For example, if team members become too passionate about a social issue, it could lead to heated debates or even physical confrontations. If conversations become too intense, it could decrease productivity as team members focus more on the discussion at hand than their work tasks. It could, in turn, lead to a decline in profits.

Therefore, employers must have clear policies and expectations when discussing social issues at work. It could include setting time limits on conversations or having designated areas where team members can discuss these topics without interrupting the office's workflow. It's also crucial for employers to provide resources for team members who may need help navigating difficult conversations or understanding different perspectives.

In other words, team members who do not care about social justice and feel that they are being forced to participate in conversations they don't want to have may become disgruntled and unproductive. They may feel that their employer infringes on their right to privacy or forces them to adopt certain beliefs. Additionally, these team members may feel that the company is wasting time and resources by discussing social issues instead of focusing on business goals. As a result, these team members may become disengaged from their work and less productive overall. For instance, they may take longer to complete tasks or make more mistakes due to their lack of motivation.

As an employer, you must create a workplace where everyone feels respected and supported. Encouraging thoughtful conversations about social issues can be a great way to foster an open dialogue within the office. Still, ensuring that these discussions are not disruptive or unfairly imposed on team members' time is crucial.

Effects on employers & executives who must face these distractions

Many employers find themselves in a difficult position regarding social justice issues. Do you know why? Because they are often tasked with finding a balance between addressing social issues and meeting the business's goals. It can be tricky, as employers must consider multiple perspectives and manage potential conflicts of interest within their teams. Also, bosses and executives may be distracted by conversations about social justice topics, leading them to neglect important tasks or focus too much on a single issue.

Executives must also be mindful of how their decisions may affect the business. For instance, if an executive chooses to focus too heavily on social justice initiatives or conversations, it could take away from their ability to manage the organization effectively.

This balancing act can be difficult for employers and executives to navigate. They may be inundated with emails, meeting requests, and phone calls from team members who want to discuss social justice issues.

Many employers are being publicly criticized for their stance on social justice issues. For example, some companies have received backlash for not taking a strong enough stance on controversial topics or for failing to address issues within their organization. This kind of criticism can be damaging and lead to further workplace disruptions.

Maintain respectful boundaries with work, personal and political issues

Now, more than ever, employers and executives must find ways to maintain respectful boundaries between work, personal, and political issues. It means understanding that while it is essential to be mindful of social justice topics in the workplace, discussions should remain professional and focused on solving issues instead of debating them. Have you ever heard of the phrase, "agree to disagree"? Well, this is a perfect example. I like to think of this principle when discussing social justice issues in the workplace.

I like to describe the case study from Coinbase, which in 2020 gave its team members who could not separate work and political/personal/social issues to leave the company with adequate severance and no hard feelings. Coinbase took a unique approach to deal with team members who could not maintain respectful boundaries between work and personal/political/woke issues. This unprecedented move highlighted the importance of maintaining respectful boundaries in the workplace.

The company allowed these team members to leave with adequate severance and no hard feelings. It demonstrated their commitment to maintaining an environment of respect and understanding while valuing diversity and freedom of thought. Coinbase took this approach to ensure team members can choose between their job and their beliefs or opinions while still expecting them to respect the company's values. This case study helps to illustrate how important it is for organizations to create a culture that respects everyone's perspectives, backgrounds, and opinions and encourages respectful dialogue. Also, it serves as a reminder of the importance of maintaining respectful boundaries between work, and personal/political/woke issues to foster an inclusive environment in the workplace. This approach helps companies avoid many potential problems from engaging in conversations that promote social justice while allowing team members to be themselves and express their opinions.

When managing these difficult conversations, employers must ensure they are respectful and considerate of their team members' time. It means thinking carefully about the subject matter of these discussions and the importance of creating an inclusive environment.

Acting on Social Justice in the Workplace

Encourage inclusive conversation

As an employer, you may already be taking steps to address social issues and create a more inclusive workplace. Some companies encourage their team members to discuss social justice issues at work and provide resources that help them engage in these conversations. For example, companies may host training sessions where team members can learn about different cultures and religious beliefs from people within their organization. They may also allow team members to take time off if they need to attend rallies or march to support a social justice cause.

However, it is essential to note that employers should refrain from forcing their team members to participate in any conversations or events that are not comfortable for their team members. By taking these steps, employers can encourage inclusive and respectful conversations that promote diversity within the workplace and bring people together.

Encourage team members to express themselves

Another way to foster a more inclusive environment in the workplace is by encouraging your team members to express themselves. It could involve creating an atmosphere that promotes open dialogue and offers opportunities for everyone to engage in conversations about social justice issues. By doing this, employers can help team members express their opinions while following their values and beliefs. To accomplish this, encourage your team members to vocalize their opinions on social justice issues whenever and wherever they feel comfortable.

Provide resources

Another way to encourage inclusive conversations is by providing team members with the tools they need to engage in them. For example, employers may provide training and resources that help team members understand the ins and outs of social justice issues and acknowledge cultural differences. They may also collaborate with other businesses in your community to bring together a group of interested people to discuss social justice work. It can help create a more inclusive workplace where everyone can learn and grow together. Is this a good idea for you?

To foster an inclusive environment in the workplace, employers need to look at their current culture and make changes that encourage respectful dialogue, community building, and open discussions in the workplace. By taking these steps, employers can create an environment where people feel safe and comfortable asking questions that challenge their beliefs. It can help them think critically about social issues while fostering new conversations promoting diversity within your organization.

Prevent tainting your reputation

Another way to help promote social justice in the workplace is by ensuring no one feels threatened or intimidated by other team members or the company itself. You may want to remind your team members that it is never okay for them to harass, threaten, or intimidate another person in the workplace. It may also involve acting against team members who create offensive comments on social media that poorly reflect your organization. I'm sure you don't want to lose all of your hard-earned reputation in one fell swoop.

I can't stress enough the importance of trusting your team members with their left (and right) brains and letting them express themselves while respecting their opinions and beliefs. If you do this, you'll find that they'll become more involved as they feel comfortable and comfortable speaking up.

Chapter 3

Motivating Engineers

Now, let's talk about the flip side of hiring engineers. Many companies hire engineering graduates to tackle tech-based products, and they hope these engineers will lead the way and translate complex requirements into simple software code. But one thing that drives many engineers to quit is a lack of motivation or stagnation.

It can be challenging to keep engineers motivated as they often have a lot of powerful ideas and yet may need to receive the recognition or rewards they deserve. Furthermore, engineers frequently feel like their work is being taken for granted or seen as just another job. Therefore, employers must find ways to motivate their engineering teams to keep them engaged and passionate about their work. Let's look at the qualities of great engineers so we can better understand what motivates them.

Strong problem-solving skills and creative mindsets

Problem-solving skills and creative mindsets are essential qualities great engineers possess, but they also lead to being well-rounded in other aspects of their lives. Being a sound engineer requires having a creative mind that can solve complex problems. It also requires thinking on your feet and quickly deciding how to tackle a problem, which means you have to be very quick at coming up with a workable solution.

Genuine engineers can solve problems and come up with creative solutions. They can look at a complicated problem and think of multiple ways to solve it, often leading to the most effective solutions. These types of engineers can also take great pride in their work and have a good sense of their abilities but are willing to learn more about the technical aspects of their job and how they can improve them. When this happens, these engineers can also take ownership of their strengths and weaknesses and learn from them

In an organization, these engineers can be an asset to the company as they are typically able to solve complex problems and make them work in a way that benefits the business. When you have engineers that demonstrate these qualities, they can be a great asset to the company.

Having a sense of responsibility and ethical values

The world is becoming increasingly digital, which means an increasingly high demand for engineers who can develop online products. In many ways, the ability to work with technology is at least as necessary as being an extraordinary engineer. Engineers who can figure out how to develop successful digital products and programs are unique and can be highly sought-after by other employers. They also tend to be the ones that others tend to respect.

Great engineers will have a genuine sense of responsibility, which means they are typically able to accept that they can't finish projects on time or according to how the company wants them done. When they cannot get the job done, they will likely admit their mistakes and take responsibility for them. It means that great engineers also have an ethical code, which means they can handle delicate situations when others may be unable to.

These engineers have a sense of responsibility and ethical standards that are truly refined. These engineers are valuable when making ethical decisions and developing solutions that meet the company's expectations.

Engineers like to work on things that matter

To an engineer, nothing is more satisfying than seeing their hard work put to good use. Whether it's developing a new app that millions will use or creating a new piece of software that will help change the world, engineers thrive on making things that matter. I have a lot to say about working on something that makes money. What matters to engineers is knowing their work is making a difference. And that's why they'll always be drawn to tasks that have the potential to make a real impact. It is also why great engineers will typically be willing to work hard when they know there's a good chance their efforts will be noticed and appreciated.

Of course, when employers hire engineers, one goal is to get a return on their investment. It means employers need their engineers to produce new products that will make the company more money by appealing to consumers. It is accomplished by hiring engineers who believe in the company's mission and finding ways to motivate them.

Most engineers are naturally curious

Curiosity is what makes a great engineer. Being a sound engineer means you can explore and understand new technologies, which means being curious is a must. It's almost impossible to be a practical engineer without having the ability to solve complex problems through curiosity and critical thinking.

Curiosity also leads to engineers taking an interest in other things in life as they want to learn more about different subjects. It is essential because it shows that these engineers are generally well-rounded individuals who can connect the things they study and their everyday lives. It can be beneficial for employers as their engineers will be more likely to find solutions to problems when they look at them from different perspectives.

Engineers enjoy working with people

Engineers love interacting with others, and being around people energizes them. They often demonstrate good communication skills, especially when working with others they respect.

If you're looking for a person who can compromise and work well with others, then a sound engineer may be the right fit for your company. Engineers are also typically more motivated to complete projects well when they feel like the people around them are supportive. No one likes working in isolation, and engineers are no different.

Engineers are detail-oriented

It is another quality that most engineers have. One of the things engineers will do when working on a project is go through the details and ensure everything is in order. While this can be frustrating for those who prefer to move quickly forward when creating something, ensuring the final product is of excellent quality is necessary. It means that most engineers prefer to take their time with things and will work tirelessly to ensure everything is perfect.

It is why engineers are typically good at completing projects on time, as they'll check off every single item on their list before moving ahead. The contentment in seeing every step done perfectly is what makes engineers happy. It leads to them taking pride in the project's quality and means they're confident about their abilities.

Engineers think outside of the box

Thinking outside the box is an essential skill for engineers. Seeing problems from different angles and creating creative solutions makes a sound engineer. It means that engineers are often the ones who can find ways to make something better or more efficient.

It is why it's crucial to give engineers enough freedom to think and explore. It can lead to some great ideas that may have been overlooked if the engineer felt too restricted or had too many limitations on their thinking. In addition, it encourages engineers to use their critical thinking skills, which can benefit their professional and personal lives.

Engineers are always looking for ways to improve

Engineers are constantly trying to learn and grow as professionals. They're rarely comfortable with the status quo and are always looking for ways to improve things. Engineers can often attend conferences, read books, or engage in other activities that will help them stay current on technology trends.

This drive to learn and improve is why engineers are so successful in their careers. They're never complacent and always looking for the next challenge. It makes them invaluable assets in any organization because of the fresh ideas and insights they bring.

Engineers are great problem solvers

It is an essential quality that engineers possess. They can take a problem and break it into smaller components so they can look at it more closely and come up with solutions. It is why engineers tend to excel in their jobs, as they're able to think of innovative solutions quickly and accurately.

This ability to think critically and solve problems goes hand in hand with engineers being great problem solvers. They can take a project and develop creative solutions to make it better, faster, or more efficient. It is why they can often be found working on complex projects, as they have the necessary skill set to evaluate the problem and come up with the best solution.

Engineer Changes the world

Engineers are often misunderstood as the repairmen of technology – fixing downed services, handling late-night password resets, and stopping hackers. They are more than simple builders and maintainers; they are the lifeblood of the modern world and its innovation. Great engineers are motivated by the opportunity to innovate and understand the next big thing that improves our lives.

These engineers constantly brainstorm new ways to improve things, and many companies even recognize them as industry experts. These individuals often play a huge role in shaping industries, which is why their contributions should be valued. If you want to attract the best engineers, you need to make these career opportunities attractive to the brightest minds in the world.

Engineers are Highly Productive

It is a trait that helps their success and allows them to be productive and valuable team members with an impact on their organization. It means they can complete projects on time, which can help your company's reputation and bottom line.

Engineers are Optimists

Engineers tend to be optimistic, and they're able to see the best in every situation. They get excited about their work, which helps them focus on positive things. It can help keep them motivated when working on a project, especially if they're surrounded by people who are negative or who don't share their passion. Being optimistic is essential in any career path, and practical engineers do it with ease and without fail.

Money isn't everything, but it's a huge factor

Not every engineer is motivated solely by money, but it's a serious motivator for team members of all types. For some years, we have noticed an expansive increase in total compensation for software engineers. The demand for technology experts, especially those who are hands-on, is increasing and will continue to do so.

Engineers can quickly break into the six-figure brackets and grow their careers by moving from company to company. First, you must ensure you are offering genuinely competitive compensation. Employees worrying about their pay are not worrying about their business outcomes. Second, evaluate your employee performance and compensation to ensure you remain competitive. Lastly, talk to your engineers regularly about their performance, compensation, and company goals.

That said, you don't have to give the farm away. I'm not telling you to pay everyone equally and cap their compensation at the market rate. You need to have regular conversations about career growth and be willing to address these issues head-on with an informed perspective based on reality.

Don't let a competitor take your talent from you because you had no idea they felt undercompensated or stagnant in their careers.

Engineers get an outstanding work-life balance

While there is certainly flexibility in every job, engineers can take advantage of this. The fact that they can create their hours, take weeks or months off, or come in early is an attractive part of the job. Engineers are happy working because they love what they do, and they genuinely enjoy coming to work every day.

Work-life balance and other non-cash benefits are excellent motivators and attractors of quality engineering talent. Once an individual's financial goals are met, they focus on benefits that improve work-life balance over additional cash. I'm one of these people.

Engineers are great communicators

It is another essential skill that engineers have that's not often found in other professions. They're not only able to convey information clearly and efficiently, but they can also effectively communicate with the team and the client. When working with engineers, it's easy to communicate your ideas, and you'll almost always get the response you want.

Engineers are usually geeks

Engineers get excited about cool new technology, which helps them do their job better as they have a deeper understanding of what's possible with technology today. It means that they can create even more innovative and creative solutions because of their background and technical understanding.

Engineers tend to be healthy

It is another trait that many engineers often share. They have a state of mind that allows them to stay focused on what they're doing, which is why they're usually very productive. It can also help them lead a healthier lifestyle, which means you'll see fewer sick days and more productivity from your engineers.

Engineers work in multiple industries

While many engineers start working in a specific field, such as design or construction, they frequently change jobs. They're always looking for opportunities to help them grow and expand their career options. A career in engineering means that you're able to work on a variety of projects, and it also means that you can work with a wide range of different people.

Engineers get to travel the world

It is another significant part of being an engineer, especially for those interested in traveling. A career in engineering allows you to take your career on the road, and you'll be able to travel to exotic and beautiful locations. It is a big reason why engineers tend to be happier and more fulfilled in their careers.

Supporting conference attendance is an excellent way to scratch the travel itch for engineering. Technology conferences are generally held in desirable central locations across the globe that provide opportunities for your engineers to grow their skills, strengthen their networks and build business relationships. In addition, these provide opportunities to generate sales leads, prospect for future team members, and drive innovation.

Engineers can work virtually anywhere

It is another attractive trait that you probably need to consider. Engineers can work virtually anywhere, meaning they have the free will to choose where they want to live and what lifestyle they want to have. It means you'll find many talented engineers who are displaced individuals who are rebuilding their lives in different places worldwide. With a career in engineering, you can take your career anywhere, yet another reason why this is a great career choice.

I once had an employee working aboard his sailboat – traveling the United States and Canada's east coast while working dutifully for our client. Except for the nautical theme to his workspace, you'd never know he was out to sea. It was one example of an employee who took advantage of his remote capabilities to build an exciting life and productive career.

Engineers have some qualities that make them great at what they do. From their detail-oriented nature and ability to think outside the box and solve problems, they bring a lot of value to any organization.

Motivating Engineers to achieve their goals and reach their full potential

Now that you know more about the qualities that make engineers great, it's essential to motivate your engineers to succeed in their roles. The following are some tips that can help you motivate your team and keep them happy at work while they achieve their career goals:

Praise your engineers

It's important to praise your engineers' efforts, especially when they do something extraordinary or above and beyond. When you praise them, it shows them how much you appreciate their hard work and motivates them to continue performing at a high level. For instance, you can commend them for completing a project or receive an award for the design of an innovative product. You can also acknowledge their work by writing them a letter or making a public announcement.

For instance, let's look at this scenario: A senior engineer in your company has developed a new piece of technology that will change how people travel. Consider this extraordinary, so you can write a letter to the engineer and thank them for their contribution. The letter should also include details about how excited you are about the technology and how it can make an impact on the lives of others.

They'll feel great when they receive praise because it shows them that you appreciate their efforts, which means the engineers will want to do more of what they're doing. Also, you can add a twist to this scenario by putting the letter in an envelope along with a gift to make things a little more interesting. Not only will your engineers receive praise, but you will also be able to show that you care about their careers and co-team members.

Let engineers work independently

Engineers are typically independent experts, meaning they like to work alone. Controlling their time and workload is essential for your engineers so that they can ensure the quality of their work. At the same time, you need to create an environment where your engineers can be creative and make decisions without being told what to do. It will allow them to take ownership of their decision-making and help them make future improvements. It means you must give your engineers the freedom to work alone and build their projects without managers' supervision. However, you can provide them with guidance and direction when necessary to achieve their overall goals and stay on track. Have you ever seen a bird fly in the sky? A bird tends to fly in its manner without any limitations. It was designed to do so. Similarly, let the engineer fly freely and see what they can achieve, yet provide them with the necessary guidance for growth and development.

When you give your engineers more freedom and control, you'll be able to make sure that they stay motivated throughout the work process. At the same time, you can help them avoid distractions and stay focused on their goals. Engineers are great because they can take ownership of their projects and make decisions based on their judgment. Giving them this freedom will encourage them to grow their career by learning new things, which is why you should let them fly.

Provide career training

Just like any other career, it's essential to give your engineers career training and help them grow their professional skills. It will motivate your engineers to continue moving up in the company, and they will achieve more in their careers. When you provide your engineers with training, they'll be more likely to perform at a high level while working. When you introduce them to new skills and train them on new technologies, it allows them to develop their expertise and become better at what they do. It's essential to give them a choice whether they want to learn additional skills or work on specific projects that interest them.

Career training is an investment because you're helping your engineers reach their full potential. Think of it this way: You'd rather have an engineer who can do a wide variety of different things instead of one who is limited. When you invest in your engineers by giving them the training they need and helping them improve their skills, you will get the most out of them in the workplace.

An excellent example is an engineer who can work on multiple tasks simultaneously. It allows them to multitask, and they will be able to work well with other departments in the company, which is why you must allow them to do so.

When you provide your engineers with career training, you'll help them push forward and reach their full potential. At the same time, you are equipped to retain top performers because they're more likely to stay motivated when given career development opportunities like this.

Celebrate milestones

Engineers are typically competitive by nature and want to achieve great things in their careers. When your engineers achieve specific milestones, you should celebrate their success with them instead of ignoring it or treating it as a regular event. Celebrating success is an art that must consider the individuals, projects, and businesses being celebrated. Consider a cash bonus if the project was accomplished by people going above and beyond (i.e., working overtime/weekends.) If it's a product of everyday business, a dinner or happy hour can go a long way by breaking out of the work routine as a reward.

Remember that it's important to celebrate your engineers because they need to be consistently recognized for their achievements, which is something you should change to help them grow within the company. I enjoy after-hours events like happy hours because you connect with other team members and spend time with them.

Set goals

Setting goals for engineers is critical to ensure the product of their effort is aligned with their career goals and business needs. These goals motivate people to focus their work on the desired outcomes without needing to micromanage. Similarly, it allows you to evaluate their skills and help them improve from your continuous feedback. In addition, if you have an objective to complete a project and reach a specific goal, you're more likely to finish it and ignore any distractions that may get in your way. Setting goals for your engineers is crucial because it ensures that their work will be consistent throughout the entire project. It also allows them to learn new things.

At this juncture, let us look at an example of how you can set clear goals for your engineers:

Set quarterly objectives that align with the company's strategy, but don't make it too complicated.

The key to setting goals is to have an idea of what you want your engineers to achieve, but don't make it too hard for them. Setting goals for your team members that are too difficult could be discouraging and cause your engineers to give up halfway through the project. On the other hand, if you make it too easy for them, there won't be many challenges, or they may not feel motivated. If you give them a goal that seems impossible at first glance and they end up conquering it, they will feel proud of themselves.

When setting a goal, think about whether you want to focus on specific projects, personal development, or the company as a whole because it's important to set long-term objectives. It would help if you also considered how this goal would benefit your business and your engineers in the future. I like to think of goals as opportunities for your engineers to grow individually and as a group. By setting clear goals, you'll be able to evaluate their skills and help them consistently improve over time.

Set up technical objectives.

The key to setting clear goals is to make them specific. I like the idea of setting objectives that focus on your engineers' engineering skills and technical knowledge. If you want to measure how well they're doing, it's also crucial to set objectives for this. By giving them specific things to work on, you'll be able to help them develop their skills and prevent them from becoming bored. Let's look at this scenario: If you have an engineer who is stuck on a specific task, you can help them set up technical objectives that will allow them to become more familiar with this task and improve their skills by breaking down the issue into smaller chunks.

There are many ways that you can measure the development of these skills. To track their progress, you should use a skill evaluation matrix (SEM) that includes technical and non-technical competencies. You should also include a section for personal growth because this is something that all engineers want to do. If you want to measure non-technical skills, you can use the job analysis questionnaires (JAQs) that will give you a way to gather feedback from your engineers. By having clear objectives for your engineers, you'll be able to track their skills and evaluate them at milestones.

Understand the risk of terminating engineers.

It's essential to understand that you'll need to terminate some of your engineers if they're not meeting your goals - it's simple math. If your engineers are not performing as expected, and you've set clear goals, you'll need to terminate them. It would help if you also considered that firing an employee can cause some damage to the company's morale. To avoid this scenario, you'll need to set clear goals for all of your engineers and with consistency.

Overall. Setting clear goals can help you and your engineers to achieve better results, so start setting clear objectives for your engineers.

Hold regular face-to-face meetings

It is critical to meet with your engineers face-to-face, in person, but virtually with cameras is sufficient. We all have different ways of doing things, but this will help you connect with them and discuss their skills and opportunities.

You can either meet in person or start the meetings via video teleconference. In person, you can have informal or formal meetings that include a presentation to your team about what you need, what scares them, and all of their recent accomplishments. It lets your engineers see that you're paying attention to their achievements and let them know where they're lacking.

I know how busy you are, but your face-to-face meetings can make a difference in how your engineers respond to your feedback. Set up these meetings because they can clarify what you want from them. If you give them clear objectives, it will help to communicate what you expect from them. At the same time, these meetings allow you to evaluate their progress and give them growth opportunities.

Demonstrate trust and respect

It's all about trust! Trust and respect are two significant factors your engineers will need to continue building the product. It means trusting them enough to know they're capable of doing the work and giving them ownership over the work.

Trust them enough to give them some creative freedom. Let's look at this scenario: When an engineer comes up with a design solution, they might need clarification as to whether or not they should tell you what it is, even if it looks different from what you expect. If you want to encourage your engineers to trust you enough, it's crucial to set clear expectations and communicate them. When your team members see that you trust them and know what they're capable of doing, they'll be more likely to share their ideas with you.

As I mentioned earlier, your engineers must have ownership over their work because this encourages them to take ownership of the process and their development within their team. If you want them to feel like they're in control of their work, you must give them some freedom and trust. You must trust them entirely with their work to avoid being a bottleneck.

Encourage collaboration and ownership

Now, you want to encourage your engineers to work on their strengths, but it's also vital that they can learn from other engineers. If you have a mix of strong and weak areas within your team, you'll need to ensure that your more vital engineers can teach their weaker engineers how to work through challenges themselves. It might seem counterintuitive, but this is part of the development process for your weaker engineers.

It allows them to build a solid foundation instead of doing everything alone.

It can be a complex process, but if your engineers can build up their skills through collaboration and ownership of the work with others, they'll be able to take more ownership of the product they're building. If you desire them to take more ownership over the product they're building, encourage them to talk about what they're doing. It will allow you, as their manager, to see where they're succeeding and where there's room for improvement. I like to describe this as a "whole team contribution" that can provide many growth opportunities.

At the same time, engineers must share work with others even if it might seem like something other than their expertise. It is important because if you want your engineers to be able to identify weaknesses and strengths within the team, sharing ideas is an excellent way for them to build on what they've already done.

An excellent example of this concept is in the modern world of DevOps (Developer Operations), a term coined to describe engineers that focus on the automation of infrastructure, software delivery pipelines, security, and operations tasks. The culture of DevOps professionals is to break down the wall between areas of expertise and share in the overall delivery of a product.

Make use of data-driven decision-making

You are probably familiar with data-driven decision-making, but I'll briefly describe it. If your team has a strong culture of collaboration and communication, using data-driven decision-making can be very beneficial.

Talking about data-driven decision-making, the key is to use the information you have within the team. An example might be that one of your engineers might want to bring his design over to another engineer because they disagree on a particular feature. If your team has a culture of data-driven decision-making, you can use this information to help you determine whether or not the change is necessary. You can better adapt to your team's current needs if it is.

While this is a straightforward example, ensure that you provide your engineers with the tools they need to use data within their work. You can measure the success of individual features and identify which ones are performing better. It will allow your team to track their work performance and identify trends in their development and overall product.

One great way to test designs, features, or improvements is with tools like feature flags, A/B testing, and individual experiences in your product or service. By implementing more than one idea or solution, deploying a sample, and measuring the performance against your business's key performance indicators, you can make decisions more confidently and foster an innovative culture.

When you communicate this information, your engineers can use it. It is an excellent way for them to track what's working and where deficiencies impede the creation of great new features.

Set up your engineers for success

Setting up your team for success builds the foundation for future success. You can help them be successful; it will allow them to become more productive and stay on track with their work. When you set things up correctly, you'll be able to identify development bottlenecks and problems earlier in the development process. It will give your team more time to identify potential solutions early on. I like to describe this: "The earlier you catch problems, the less time they'll force you to waste fixing them."

As you can imagine, setting your team up for success involves documentation and communication. The major factor in determining whether or not your engineers are successful in their interactions with their teammates. If one of your engineers wants help from another engineer and needs help understanding what he's trying to accomplish, it can become problematic. You want to provide your engineers with clear and concise information so all your team members can understand precisely what's going on and why. Additionally, you have to ensure that this information is communicated accurately to ensure everyone can make the best decisions possible. Engineers who need to understand why something is being done might encounter problems later.

Acknowledge mistakes as learning opportunities

You want your engineers to feel like they're in charge of their work, but you should still pay attention to mistakes and flaws. Instead of punishing your engineers for mistakes, you should use these problems as learning opportunities.

Let's say an engineer has identified a problem with one of the features in the app. Maybe he's made a mistake, or something about the product needs to be fixed. Instead of yelling at him, you want to acknowledge the issue and help him to understand how he can fix it. Once he learns how to fix it, he'll be able to work on the next issue and make sure that your customers have a great experience using your product.

Establish a performance management system

Performance management is something you're familiar with, and it's an integral part of the development process because it helps you identify some of your team's weaknesses. At the same time, it allows your engineers to succeed and build on the strengths that they already have.

Let us talk about performance management; you must establish a system that works for your team. My favorite system for performance management is what I like to call "the three As." Here is how it works:

The first A is about accomplishments, which is a great way to identify the new features that your engineers have built. If you have a website that provides a tracking system for the new features, having this information will allow your customers to understand better what's working for them. If an engineer has identified room for improvement and successfully built something different, you must provide him with positive feedback. It will encourage him to develop his skills and build on his strengths.

The second A is about action, allowing you to identify how your engineers work together with their teammates. As mentioned in step six, your engineers must work together within the team and provide one another with positive feedback. The achievability of goals is something your managers will be responsible for identifying. They must focus on what's working and use their experience to help the team grow.

The final A is about attitude. It is all about your engineers' attitude whenever they're working together. If a group of your engineers isn't getting along, you must analyze their relationship. Are they constantly arguing with one another? Or can they work together and provide the level of positive feedback needed to motivate each other? If you notice problems within their work processes, it might be time to re-evaluate how you work together as a team. Hopefully, you'll grow and provide a better environment for your engineers to succeed.

Performance management is vital for the development process of your product. It provides a great way to identify any problems or deficiencies within your team, so you want to establish them as soon as possible.

Leverage technology to its full potential

You know the importance of developers, designers, and administrators in product development. However, an additional layer within the development team must be considered: the environment.

As discussed in steps one and two, you want to ensure that your engineers can work together as a unit. As a result, they must have the right tools available so they can accomplish their tasks. Fortunately, technology has made this process more accessible than ever before.

For example, you might have a great engineer who's very knowledgeable in a specific technology. It might have helped him develop some of the best features in the app, but there may be other engineers who need to learn more about it. If your organization doesn't provide access to this technology, it could cause some problems. If one of your developers has access to these resources, he'll be more likely to stay motivated.

You must provide your engineers with the technology they need to accomplish their goals. If it's an expertise that benefits all your team members, then it's something you should take advantage of and utilize. Otherwise, you might miss great opportunities because you need the right technology.

Make the work environment comfortable and enjoyable

A comfortable environment is something that everyone should want. After all, it's one of the main reasons why workers continue to work for their companies. The best part? You don't need a study that proves this because you can easily measure with your own eyes.

Whenever your engineers are working on the development process of your product, they should feel like they're in a joyful environment. It means making your team members feel like they're valuable assets to the company by providing them with tools and resources that help them achieve their goals.

My favorite quote is: "*If you want people to do a great job, give them a great environment.*" You must provide your team members with the environment they need to succeed. It could be different for every person, so it's something that you'll need to evaluate on an individual basis. Engineers will be working with various technologies, so you must provide them with comfortable chairs, desks, and workspaces. The same thing goes for their work environment. If their work environment is uncomfortable, your team members will be less likely to put in the effort needed to accomplish their goals.

I've always been a firm believer in an open office space. This type of environment allows your team members to have access to one another, allowing them to help each other out whenever a problem arises. It also encourages transparency and collaboration, which is what you want from your engineers.

Chapter 4

Qualities Engineering

Leaders Should Possess

Now that you understand how to create an engineering culture that motivates and encourages team members, it's important to remember the need for leadership with purpose. A great leader will be able to motivate their team members and possess several qualities that make them a valuable part of the organization. *Let's look at the various qualities a leader should possess.*

A leader knows that they are responsible for their team members

Engineering Managers, Executives, and other leaders will be responsible for their team members' success. When I say responsible, I mean that they are the ones who will be responsible for the job performance of their team members. It means they need to be involved in everything that goes on within their organization. They must know where each person stands, including work problems or achievements.

It doesn't mean that a leader will be micromanaging every aspect of every person's job performance. It isn't something you want to do as a leader. Instead, what you should be doing is making sure that a person understands their role within the company and how they can be successful. In a way, you will serve as a motivational coach since you're providing your team members with the support they need to be a successful part of your organization. To do this, you want to focus on your company's mission and create a plan to help each person accomplish their goals.

You must avoid getting too involved in the day-to-day activities of your team members. It can lead to micromanaging, making it harder for engineers to have an impact. By keeping yourself out of the details, you can focus on providing strategic direction for your team members and their success. Here is a practical scenario of what I'm talking about:

Let's say you have a team member having trouble with a particular project. Instead of jumping in and trying to solve the problem yourself, talk to your team members about what they need to do differently or if there is something specific they are struggling to get right. It will allow them to take ownership of their success while still providing them with the guidance they need to succeed.

A leader knows how to prioritize

It would help if you were always busy taking on a challenging project or helping a team member when needed. However, as a manager or executive, you must also be aware of the most important tasks and focus your time and energy on those. Prioritizing is an integral part of being a leader, and you need to be aware of if you want your team members to stay motivated.

Prioritizing means identifying what tasks will have the most impact on the success of your organization and making sure that those are taken care of first. It means that you need to know how to delegate tasks to your team members so they can stay focused on their work while still being able to help out with essential projects. I like to ensure that your team members run in the same direction.

By taking the time to prioritize, you're able to create a more efficient work environment that encourages everyone to stay motivated and ensure they're reaching their goals. An example of how this is beneficial is illustrated here: Let's say you have two projects that must be completed by the end of the week. Instead of having both teams work independently, try combining them and assigning tasks based on each person's skill set. This way, you'll be able to get both projects done faster while also ensuring that everyone has something they can work on that is meaningful and challenging.

By taking the time to prioritize, you're creating an environment where everyone feels like they're part of something bigger and more critical than their tasks. It will help inspire your team members and ensure that they are motivated and stay on track with their goals.

You might wonder how to become a better leader and motivate your team members. Remember, leadership is more than just delegating tasks; it's also about setting an example and providing the proper support. By taking the time to prioritize, think strategically, and provide meaningful guidance, you'll create an environment where everyone feels inspired and motivated.

A leader makes sure everyone is heard

To create an inspiring work environment, it's crucial that every team member feels like they are being heard. It means listening to their ideas and ensuring their opinions and contributions are considered in any decision-making process. Everyone wants to feel like they matter and have an opinion on what goes on in their workplace.

It does not signify that you have to agree with every idea or suggestion. Still, it does mean that you should actively listen and consider every team member's opinion before deciding. This kind of active listening helps build trust between you and your team members and shows them that you value their input and are taking the time to understand what they're saying.

As a leader, it's essential to ensure everyone is heard and respected. If team members feel like their opinions aren't being taken seriously or are not being listened to, they won't be motivated to put in the extra effort necessary for success. By actively listening and considering every team member's opinion, you'll create an environment where everyone feels included and valued—essential for any thriving engineering culture.

Creating an engineering culture that motivates your team takes work and dedication from management and team members. But by having a purposeful approach toward leadership, prioritizing tasks carefully, delegating strategically, and making sure everyone is heard, you can build an engineering culture that will keep your team motivated for years to come. You desire to be successful in your engineering efforts, so make sure you take the time to cultivate a culture of motivation.

A leader knows how to manage conflicts

While this is done only sometimes, a leader needs to manage conflicts among team members. Conflict can often lead to a lack of motivation and decreased productivity, so leaders need to recognize signs of conflict early on and intervene before things get out of hand.

As a leader, you should also know how to mediate disagreements fairly and objectively while keeping the team's best interests in mind. It means being willing to listen to both sides without bias, asking probing questions, and having an open mind when considering solutions. Ultimately, you must be patient and understanding with your teams while still holding them accountable for their actions. I describe it as the leader with an "iron fist in a velvet glove." It means maintaining a robust and good presence with your team members to build an atmosphere where everyone is respected and heard. Here is a detailed scenario of what I'm trying to say so this doesn't sound too abstract.

For example, two of your team members disagree about a task. It would help if you started by asking each person to explain their side of the story and then work towards a compromise between both parties. It could involve each person compromising some of their ideas or working together to devise an alternate solution. Whatever the outcome may be, it's vital that everyone feels heard, respected, and understood before moving forward.

By learning how to manage conflicts among team members successfully, you will create a culture where people feel comfortable expressing themselves without fear of judgment or retaliation. When this kind of atmosphere is cultivated, it leads to increased motivation and productivity from your team. The IT space is full of changing tech and strong personalities, so having a leader that knows how to navigate these tricky waters is essential.

A practical engineering culture needs leadership with a purpose. Ensure you're actively listening to your team members and valuing their opinions while managing conflicts objectively and fairly. Doing this can create an environment where everyone feels motivated to work hard and achieve success.

A leader sets realistic expectations

Now, let's talk about setting realistic expectations. Leaders must be honest about the scope of a project and the timeline that should be adhered to for it to be successful. Setting unrealistic expectations can lead to frustration and burnout from your team members, so it's essential to consider their workload when assigning tasks. Think of it this way: if you make your team members promise to finish something in an impossible amount of time, they will be less likely to stay motivated.

Similarly, leaders must provide ongoing feedback and support throughout the project timeline. It can help ensure that goals are met, and expectations are realistic. I see it as having a "check-in" with my team regularly. It can involve asking questions about progress, offering support for any problems that may arise, and providing helpful resources to get stuck tasks back on track.

Leaders must set realistic expectations and provide ongoing feedback so their team members stay motivated and focused. You may ask, "How do I know if I'm setting realistic goals?" The answer is simple, and I want to answer it with the "3-C" rule. Let's break it down:

The 3-C rule stands for Communication, Cooperation, and Coaching.

Communication: Communication is the foundation for setting realistic expectations. Ask questions and listen to your team members when they express their thoughts about tasks. It will help you understand how much time a task may take, what challenges may arise, and what resources may be needed to complete it. It also helps create an environment where everyone feels comfortable expressing their opinions and ideas without fear of judgment.

When you listen to your team members and use this feedback to adjust expectations, you set a tone of respect and understanding that can lead to more motivated team members. For instance, if a task seems too daunting for one person to tackle, try breaking it up into smaller pieces and assigning parts of the task to different team members. This way, nobody feels overwhelmed or burned out.

The IT space is filled with challenging projects and people that need to be managed. So, as a leader, it's crucial to create a culture where communication is encouraged between team members and expectations are set realistically.

Cooperation: Cooperation helps ensure that tasks are completed efficiently. Work with your team members to devise creative solutions to problems and delegate tasks in a way that makes sense for everyone. Foster an environment where team members can offer ideas and assistance without feeling like their input isn't valued. It will ensure that everyone feels invested in achieving the same goal. I think of it as "we are in this together." It can motivate team members to stay focused and complete their tasks on time. For instance, if a task requires multiple people to complete, assign roles according to each team member's skill set. This way, everyone works together towards the same goal in the most effective manner possible.

Finally, coaching is essential for long-term success; it helps build trust and loyalty among team members by showing them that their efforts are valued. A leader should take the time to mentor and coach individuals on specific tasks or projects rather than simply giving orders and expecting them to be followed. This type of guidance and support helps team members grow while motivating them to stay focused and overcome any obstacles they may face.

Technical Expertise

A solid technical background is a must for any engineering manager, and staying up-to-date with the latest technology trends is essential. This technical expertise enables you to understand and make decisions about different technologies and provides direction for your engineering team. I'm sure you already have some experience in the IT space, but it always helps to brush up on your knowledge. Take some time to read industry publications and talk with colleagues who are more experienced than you.

By staying informed about current technologies and trends, you'll better understand your team's challenges and provide helpful solutions. You want to stay caught up because your team would lose motivation quickly.

Combining these elements creates a culture where engineering teams are motivated to strive for excellence and feel appreciated for their hard work. Leaders who take the time to create an environment with meaningful conversations, cooperation between team members, and development coaching will always have a successful team in the long run.

One way of combining these elements into an effective motivator for your team is to form Technical Guilds within your engineering team. Come together regularly to discuss the latest industry news or trends; explore new technology, tool, or service; or put heads together studying for a challenging certification exam.

Relationships

You're going to need help to lead and motivate a team. Each team needs a leader; this is where the importance of relationships comes into play. As an engineering manager, you should build lasting relationships with your team members. Building relationships is not only crucial for making sure everyone is working together, but it also creates transparency among the team.

Here are some guidelines to follow:

Make a deliberate effort. Sure, it may be awkward at first, but you have to make an effort to build relationships within your team. Think about everything you will be asking them to do for you and ensure that they know that you care about their success and well-being. It could be by ensuring that they:

Understand their capabilities. You can identify a team member's strengths and weaknesses by building relationships. It will help you understand the tasks they are most capable of completing on their own, and it will also help to identify any challenges that may arise during their work. It is huge!

Feel appreciated for their work. Finally, as I've said earlier, one of the essential things you can do for your team members is to make them feel appreciated for their work.

Keeping a business relationship professional by following these guidelines will help you set boundaries in the relationship – preventing it from becoming fraternization – which has long-reaching negative implications in larger teams.

Professional Networking

When starting as an engineering manager, you will likely need more money to hire a public relations person to put your company out there. And if you get that luxury, I still highly recommend networking on your own!

When you're networking, it's essential to make sure that you don't just talk about yourself and your company but build relationships with the people in your network.

It is how you ensure that people see value in the conversations. Take them out to coffee or lunch, ask them what kind of projects they're working on (or hiring for), and determine if there is any way you can help. By genuinely being interested in what they're working on, you're creating relationships and increasing the potential for collaboration.

Being able to network in a professional setting will also open up new opportunities for your growth as an engineering manager. As your network grows, you increase your chances of discovering new job opportunities or getting valuable advice from experienced professionals. It is a beautiful way to stay up-to-date with industry trends and ultimately build out an impressive portfolio of work.

The more people know about your work, the more doors can open for you! So, get out there and start building relationships! You do not know where it will take you.

Change Management

The IT industry is constantly changing, and engineering teams need to be able to adapt quickly to stay ahead of the game. As an engineering manager, technology executive, or head of an engineering team, it's essential to lead your team through any changes that may come their way.

Change management is a process in which an organization looks at the impact of change on its team members, processes, and systems. This process has four critical steps: 1) Assess the Impact; 2) Plan for Change; 3) Implement the Change; and 4) Monitor the Change.

Let's look closely into each of these steps.

Assess the Impact: It is the most vital step of the process. It involves looking at how organizational changes will affect your people, processes, technology, and any other aspects of the business. Ask yourself: Is this change going to support (or inhibit) achieving our business goals?

Plan for Change: After assessing the impact of change, it's essential to put a plan in place for how to roll out the changes. It would help if you involved your engineering team and other stakeholders who the new change may impact. An effective communication plan will ensure that everyone involved understands the process and timeline of implementing the change. Ask yourself: Is any part of this plan ambiguous?

Implement the Change: Once the plan has been finalized, it's time to start implementing it. It is where a strong leader comes in - someone who can keep everyone on track and quickly address any questions or issues.

Monitor the Change: Finally, it's crucial to monitor the progress of the change over time to ensure that it is successful. It includes looking at the impact on team members, processes, and technology, as well as feedback from stakeholders. A data-driven approach is needed here to prevent internal bias from impacting the outcomes. Ask yourself: What specific, measurable, and relevant metric can be used to determine success or indicate failure?

When implementing changes, it's essential to recognize that different people react differently to change. Some may look forward to it, while others may struggle with it. As the leader, it is up to you to create an environment where everyone can feel comfortable with the changes and be motivated to work together toward a successful outcome.

Dealing with introducing change with your engineering team is an art form. One method that I often have success with is idea seeding. Essentially, propose your idea in the form of a problem – offer background information, a user story, and some criteria for success. Guide your team through the solutioning process, nudging them towards the desired change. By doing so, the team has the illusion of being part of the solution. Your team members are more likely to be positively engaged in its execution. Alternatively, on a wrong path, the team might highlight some problems with the change and guide YOU towards a better solution.

Overall, change management is essential for any engineering manager to learn. With practical change management skills, you can ensure that your team can make smooth transitions when changes are needed to stay ahead of the game. So, get out there and start leading with a purpose!

Data Analysis and Interpretation Skills

As a leader in an engineering team, you should be able to analyze the data from past projects and use it to make informed decisions about upcoming initiatives. It includes understanding how your team's performance is measured and how the results are interpreted. To do this effectively, you'll need to develop strong data analysis skills, such as skimming charts and tables, identifying data patterns, and drawing meaningful conclusions from the results. Additionally, you should communicate your findings to other stakeholders and use the insights to create strategies and plans for future endeavors. With robust data analysis and interpretation skills, you are better equipped to allocate resources and budgets appropriately.

Effective use of data in decision-making is a two-way street. In one way, it's a scientific approach to determining risks, understanding your levers and knobs, and providing evidence to support ideas or change. In another way, everyone interprets data differently based on their perspective, which may contradict your desired outcome. A strategy that has been successful for me is to present data, charts, diagrams, or other supporting materials with the least amount of information required to make your point – while also making your point clear.

A simple way to accomplish this is to create a headline to make your point – the headline should be short, to the point, and describe the action you expect to take. For example, "SALES PIPELINE UP 50% QOQ, INCREASING ENGINEER HIRING FORECAST." It is suitable to the point and ignores any specifics, such as close rate, lead qualification, or probability, that may cast doubt on your proposed decision. When done in good faith, these builds trust and allow stakeholders to open a conversation (or not) with additional details.

Relationship Building

Finally, relationship building is an essential skill for engineering leaders. After all, you'll work with various stakeholders with different interests and objectives. It's vital to establish strong relationships between team members and other stakeholders. Think of it this way: when everyone involved in a project understands each other's needs and perspectives, it will be much easier to develop effective strategies and reach successful outcomes. To become a great communication leader, practice active listening skills, learn how to build trust between yourself and the people you work with, and always strive to create an open and collaborative environment.

Let's look at each of these points in more detail:

Engage in listening actively. Active listening is a skill that can help you become a better leader, as it allows you to understand what others are saying and ensure everyone on the team is on the same page. To practice active listening, pay attention to the other person and avoid interrupting. Additionally, ask open-ended questions to get more information if needed.

It is imperative to get active listening right. It is deliberately focused, and your peers will know when you're distracted with Slack, Email, your phone, or your cuddly cat. It's best not to try to get away with it – if you're not needed 100% in a conversation, leave it.

Learn to build trust. Building relationships of trust is essential for successful teams, as it allows team members and stakeholders to work together effectively and efficiently. To do this, focus on developing open communication channels, be honest about your thoughts and feelings, and always strive to act in the team's best interests. A great leader will also empower team members to take ownership of their tasks and be available to provide help and guidance when needed.

Building trust by discussing your feelings can be a powerful tool in conflict resolution or negotiation because they are inarguable and potent. Use them to your advantage, we're all people, and we all have emotions that come into play at work.

Create an open and collaborative environment. A vital part of a thriving engineering culture is creating an open and collaborative environment where everyone feels comfortable sharing ideas, asking questions, and working together towards common goals. As I've mentioned, active listening, trust building, and open communication are essential. Additionally, ensure a safe space where team members feel comfortable speaking up about any challenges or issues they face. I can't stress these three points enough.

Understanding different leaders

A great engineering leader understands the importance of cultivating an environment where teams can work together effectively, focus on developing strong relationships between team members and other stakeholders, and employ data analysis and interpretation skills to make informed decisions. I like to break the types of leadership down into the following categories:

1. Assertive Leaders – (People who give it to you straight)

Have you heard the phrase, give it to me straight? As a leader, it can be in the form of a manager asking their teammates or team members whether they did something right or wrong. Also, it can be a co-worker asking another co-worker what they think about the new company policy.

An assertive leader gets to the point when giving feedback and making suggestions. They are direct and to the point and don't beat around the bush. These people are usually the type who is used to giving feedback and offering solutions right away. And they aren't afraid to say no when necessary, either. The assertive leader is confident in their ability to solve problems and accomplish the job. He is typically a firm believer that everyone deserves a fair representation in their team regardless of their background.

Let's look at the common attributes of assertive leaders:

• **Clear communication**: Assertive leaders are clear and direct with their communication. They aren't afraid to speak up and express their opinion, even if it's in opposition to others.

• **Strong decision-making skills**: Assertive leaders can make tough decisions quickly. They trust their gut instinct and have the confidence to move ahead with decisions even when there is dissent from their team.

• **Respect for all opinions**: Assertive leaders are open to hearing different opinions and viewpoints and can make vital decisions based on these opinions. They don't ignore potential solutions or ideas as they know every perspective has value.

• **Wastes no time**: Assertive leaders don't waste time getting the job done. They understand the importance of working fast and efficiently and can move ahead on tasks without any delays.

How can these attributes add value to you as an engineering manager or IT executive?

Assertive leaders bring a sense of urgency and direction when getting the job done. They understand the importance of moving quickly and efficiently towards a common goal and trust their decision-making abilities to get the job done. This type of leadership is essential for any business, engineering or not, that needs to deliver products or services intently. When managed well, assertive leaders can be an asset for any engineering team. Here is a scenario to consider:

You are the head of an engineering team and have been tasked with developing a new product. Your team is behind schedule, and everyone is stressed about meeting the deadline. You have an assertive leader on your team who can step up and take charge when it is needed. This leader provides clear direction, makes decisions quickly, and is confident to move ahead with tasks even when there is dissent from the team. With an assertive leader on your team, you can move much quicker and efficiently towards meeting the deadline.

As you can see, assertive leaders are invaluable assets for any engineering team. They provide clear direction and focus the team on completing the job quickly and efficiently. If you want to cultivate an engineering culture that is motivated and productive, consider bringing on assertive leaders as part of your team or strive to be one yourself.

The dangers of being an unassertive leader

Now, you might wonder what the dangers of being an unassertive leader are. An unassertive leader tends to let decisions drag on or is reluctant to decide. This type of leadership can lead to confusion, stagnation, and even hostility within a team as everyone is waiting for direction from the leader that never comes. I'm sure no one wants that for their team.

An unassertive leader can be detrimental to an engineering team. These leaders avoid giving feedback or offering solutions due to fear of conflict or rejection. Unassertive leaders also lack confidence in their decision-making abilities and may struggle with making tough decisions when necessary. As a result, they often procrastinate or pass decisions on to other team members.

In my experience, unassertive leaders often make decisions based on committees or democracy. As a leader, it is a slippery slope, and I consider it a danger to the success of a competitive business. Coach these people out of this mentality without delay.

Unassertive leaders often get left behind, and their trust in their team members or peers is likely to be lost. Unassertive leaders may not be heard by their team members. It can have a significant effect on their success. Therefore, as a leader, being assertive is essential to gaining respect and trust from those around you. With the right amount of confidence, you can ensure that your opinions are heard and respected. Taking charge of situations, communicating effectively, and standing up for yourself will go a long way toward becoming an effective leader.

2. Mentoring Leaders – (People who lead with empathy)

The mentoring leader is a different type of leader. A mentoring leader focuses on developing relationships between colleagues and providing guidance to their team. These people listen before speaking, provide direction without being too critical or judgmental, and focus on helping each individual achieve success.

It's essential to recognize that a mentoring leader doesn't just tell people what they need to do but helps guide them in the right direction. They understand that everyone needs support and guidance at times and aim to provide this in an encouraging and understanding way. A great example is someone who takes the time to explain a concept or process to their team in an approachable way.

Let's look at the common attributes of mentoring leaders

Empathy and understanding: A mentoring leader can relate to those on their team. They take the time to listen, understand, and even put themselves in someone else's shoes.

Patience: Mentoring leaders know that everyone learns differently and at their own pace. They're patient when teaching new concepts or processes and can explain things multiple times if needed

Encouragement: A mentoring leader always looks for ways to encourage and motivate their team. They recognize successes, provide positive reinforcement, and always look for ways to build morale.

How can these attributes add value to you as an engineering manager or IT executive?

Mentoring leaders is invaluable for any engineering team. They provide essential guidance and support that helps your team face the challenges of working in a fast-paced, competitive environment. Having a mentoring leader on your team can encourage collaboration and open communication and help foster a culture of trust and respect. It can increase morale, productivity, and loyalty to the team and business. To cultivate a motivated and productive engineering team, consider bringing on mentoring leaders or striving to lead with empathy.

Dangers of not being a mentoring leader

The dangers of not being a mentoring leader are apparent. Without empathy, guidance, and support from leadership, teams can become disengaged and unmotivated. They may need more direction to succeed and be less likely to take the initiative or develop innovative ideas. Unmotivated teams can lead to decreased performance, low morale, and a disorganized workplace. The overall effect on the team and business can be tremendous, so it's essential to ensure you are taking the time to mentor, guide, and provide encouragement to your team.

Ultimately, having the right mixture of assertive and empathetic leadership can positively impact any engineering culture. As an engineering manager or IT executive, it's essential to have a team of leaders who can motivate and inspire the people around them. With exemplary leadership, your engineering teams will be motivated and productive.

3. Visionary Leaders (People who aim high)

Visionary leaders are the people that set the bar high and have a clear idea of where they want their team to go. These leaders can think strategically and see the big picture when achieving their goals. Visionary leaders challenge the status quo, look for innovative solutions to problems, and strive to go further than anyone else in their industry.

In cultivating an engineering culture that is inspired and motivated, visionary leaders are invaluable. They can help shape a team's direction by inspiring them with their ideas and pushing them to think outside the box. They can keep everything in perspective, focus on achieving long-term goals, and develop strategies to help their team succeed.

Let's look at some of the common attributes of visionary leaders

• **Strategic Thinking**: A visionary leader clearly understands where they want their team to go and can devise a plan to get there. They can think ahead, anticipate potential roadblocks, and develop strategies to overcome them.

• **Innovation**: Visionary leaders are not scared to take risks and try new things. They strive for excellence in their work and always look for ways to optimize the team's performance.

• **Empowerment**: Visionary leaders understand the importance of empowering their team to reach their goals. They give their team the support and encouragement needed to take ownership of their work, develop new ideas, and strive for excellence.

How can these attributes add value to you as an engineering manager or IT executive?

Being a visionary leader can help you cultivate an engineering culture that is driven and motivated. Visionary leaders can bring invaluable insight and expertise to any engineering team or IT executive role. The ability to think strategically and develop innovative solutions is essential for any team looking to stay ahead of the competition. By having a visionary leader on board, you will have someone who can provide direction and motivate the team to reach its goals.

Dangers of not being a visionary leader

The dangers of not being a visionary leader are apparent. Teams can become stagnant and lose motivation without someone to look ahead and anticipate potential problems. They may need more direction to succeed and be less likely to take the initiative or develop innovative ideas. Have you ever felt like your engineering team was stuck in a rut and unable to move forward? If so, it could be due to a lack of visionary leadership. As a leader, you want to ensure that your team is constantly striving for excellence and pushing the boundaries of their work.

4. Polite Leaders – (People who kill you with kindness)

How do you interpret the term "polite leaders"? Polite leaders are the people who figuratively kill you with kindness. They have an empathetic approach and strive to understand the needs of their team members. They know how to connect with each individual to create a supportive environment.

Polite leaders are the people who make everyone around them feel comfortable, respected, and valued. They have a way of speaking that creates an open and friendly environment. Polite leaders understand how to be assertive without becoming aggressive, which is essential for any team looking to create an engineering culture that motivates its members. If you've encountered a polite leader, you know how energizing and refreshing it can be.

Polite leaders have many positive qualities that make them great engineering managers or technology executives. Let's look at some of the common attributes of polite leaders:

Respect: Polite leaders show respect to everyone they work with – from their team members to their customers. They understand the importance of treating people with kindness and not taking advantage of them.

Open-mindedness: Polite leaders are open to new ideas. They are willing to listen and consider other people's points of view, which can help create a more collaborative environment within the engineering team.

Flexibility: Polite leaders can adjust their approach when needed. They understand that the needs of a team or situation can change quickly and are willing to adapt accordingly.

Negotiable: Polite leaders are the type of people who want to find a win-win solution. They do not always expect things to go their way, and they are willing to negotiate to reach a mutually beneficial compromise for everyone involved.

How can these attributes add value to you as an engineering manager or IT executive?

Polite leadership is vital for creating an engineering culture that motivates and empowers its members. Polite leaders ensure their team feels respected, valued, and appreciated – all of which are essential for high performance. By being a polite leader or having one on board, you can ensure that your team has the support and encouragement they need to reach their goals.

Dangers of not being a polite leader

The dangers of not being a polite leader are clear – it leads to a hostile work environment, which can harm morale and lead to poor performance. People who feel disrespected or undervalued are less likely to be engaged and motivated. As a leader, you want to ensure everyone feels comfortable and respected to create an engineering team loyal to your business.

IT is a competitive industry, so having a polite and empathetic leader on board can make all the difference. When you have someone who can foster an environment of respect and understanding within your team, they are likely to give their best and reach excellence.

5. Creative Leaders – (They don't suffer fools)

You often hear "creative" and associate it with "out-of-the-box thinking." Creative leaders are open-minded and don't suffer fools. They are not afraid to express their minds and explore new ideas.

Creative leaders are like a sponge in their thinking – they absorb everything around them and use it to develop innovative solutions. They view anything as an opportunity for growth and exploration, resulting in novel ideas that can benefit the engineering team or business.

Creative leaders often inspire their teams by setting a high standard of excellence and encouraging them to think outside the box. This type of leadership is essential for any engineering team that wants to remain ahead of the curve and be innovative. When you have a creative leader in place, you can be sure that your team will have the creative spark needed to create something great.

Common traits of creative leaders

Open-mindedness: Creative leaders, open-mindedness is essential. They are open to trying new things and view failure as an opportunity for growth. For instance, they will be bold, take calculated risks, and explore new ideas.

Passion: Creative leaders are passionate about their work and often have an inexhaustible energy source. They are excited to see their team succeed, which can inspire the whole team to do better.

Vision: Creative leaders have a clear vision for their team and can communicate it effectively. They know their ultimate goal, which helps them stay focused on the task at hand.

Inspiration: Creative leaders inspire their teams by setting high standards and motivating them to reach for excellence. It can help create a culture of innovation essential for any successful engineering team.

How can these creative attributes add value to you?

Creative leaders are essential for engineering teams that want to remain at the top of innovation. They can provide new ideas, inspire their team to reach excellence, and foster a culture of open-mindedness and innovation. This type of leadership is vital for businesses that desire to remain competitive and thrive in the ever-evolving world of technology.

These creative attributes can add immense value to any engineering team and help them reach their goals faster and more effectively. Therefore, if you are looking for a way to get the best in your team, being creative is the way to go!

Chapter 5

The Intellect / Emotional

Quotient

Now, having discussed the importance of creative leadership, let's explore another critical aspect of engineering team dynamics – emotional intelligence.

What is EQ?

Emotional Intelligence, colloquially measured as the Emotional Quotient (EQ), refers to an individual's ability to recognize and understand their emotions and that of others. It is crucial to develop strong relationships between team members and create a successful team dynamic. There are three main dimensions of emotional intelligence:

Self-awareness: Individuals with an excellent EQ are emotionally intelligent because they have a high level of self-awareness. They can recognize their emotions and understand how they influence their behavior and reactions. As a result, they are better at managing their own and others' emotions, which can help them cope with stressful situations. Here is a scenario to help you understand:

You work in an IT company, and one of your co-workers is having a hard time. He is constantly snapping at everyone around him, and you can tell he's struggling to keep it together. With emotional intelligence, you could recognize what was going on with him. Instead of getting angry or frustrated, you could offer him genuine support and understanding. It can help him get back on track without resorting to negativity or outbursts.

Emotional Regulation: Individuals with an excellent EQ can regulate their emotions and behaviors when faced with difficult situations. They know how to calm themselves down during moments of stress and can recognize when their reactions are inappropriate.

For example, imagine you arguing with a colleague, and things start to get heated. You can feel your temper rising and your face red, but then you take a few deep breaths, reminding yourself of the importance of remaining calm. With emotional intelligence, you know that it is the proper way to handle the situation and will help you resolve the disagreement peacefully.

Empathy: Individuals with an excellent EQ can recognize, understand and appreciate the feelings of others. They can put themselves in other peoples' shoes, allowing them to be more compassionate and understanding towards their colleagues.

I like to think of EQ as adaptability. When someone has high emotional intelligence, they also tend to be adaptable. They are less likely to be driven by their emotions – they are "checked at the door," so to say. Typically, these people are easygoing, have managed egos, and can disagree and commit.

By cultivating emotional intelligence within your organization, you can create an environment that is more understanding, supportive, and motivating for everyone involved. Everyone can benefit from having the skills to recognize and regulate their emotions and that of others. When team members feel heard and respected, they are more likely to feel motivated and loyal to their company. It can increase productivity, creativity, and morale throughout the organization. Leadership with a purpose starts by cultivating an emotionally intelligent engineering culture that motivates team members!

What is IQ?

Now that we understand emotional intelligence, let's look at cognitive abilities, measured as the Intelligence Quotient. It measures problem-solving skills, critical thinking, information processing speed, and the ability to reason logically. These skills are fundamental in many aspects of life and work, but they don't necessarily determine your success as a leader or manager.

Leaders with IQ understand the importance of making logical decisions and using problem-solving skills to develop creative solutions. For instance, if a company is facing financial problems, an IQ leader will be able to analyze data quickly and develop strategies that help the business succeed.

Leaders also need EQ to understand their teams and foster strong relationships. When leaders have high emotional intelligence, they are often better at recognizing and responding to the needs of their team members. They can motivate team members, build trust, and foster collaboration. A leader with only IQ is like a ship without a rudder, while a leader with IQ and EQ is like a ship with a captain. The leader knows how to navigate the waters. This analogy is especially true when leading a team of engineers.

There are a few main dimensions of IQ that help leaders to be successful:

Reasoning. You can comprehend and analyze data. If you have high reasoning, you will be able to understand numbers quickly, solve problems from multiple perspectives, and solve them effectively. Leaders with high reasoning skills can quickly identify the best way to solve a problem.

Analysis. It is similar to reasoning, but you will use it to analyze information instead of problem-solving. Leaders with high analysis skills will be able to look at data and determine what they need to do to fix it. They will be able to look at a situation and find logical solutions. Additionally, they can look at data-driven decisions and effectively determine their credibility.

As discussed in Daniel J. Levitin's "The Field Guide to Lies," data, charts and graphs can be manipulated or obfuscated to hide the truth and make inaccurate conclusions. These scenarios would not easily fool a person with higher reasoning and analysis skills.

Objectivity. Leaders with high IQ will be able to think objectively, which means they can understand a situation from multiple perspectives. It can also help them feel less emotionally attached and more open and receptive when working with others. I'm guessing you've worked on a project for a long time, poured in your best effort, and burned the midnight oil to make it awesome – to have it scrapped? Approaching this common scenario with pure objectivity is a difficult but essential quality of a technology executive.

Information Processing Speed. It is how quickly you can process information and make decisions. If you have high information processing speed, you can quickly take in large amounts of data and compare them -- such as comparing different sets of numbers, graphs, or charts -- quickly. You will make decisions based on those results quickly and efficiently.

Leaders with high information processing speed can quickly analyze data and develop practical solutions.

Working in the engineering field requires an understanding of complex concepts and the ability to think analytically. It also requires a certain level of emotional intelligence, such as collaborating with others, being open to feedback and criticism, and recognizing how their actions affect those around them. High EQ engineers can recognize these things and use them as fuel for success in their careers. I like to think of them as "people-smart engineers."

High IQ and EQ engineers are some of the highest-performing and thriving in the industry. They understand that the most effective teams are those that include both emotional intelligence and technical skills. By combining these two traits, they can lead with empathy, think outside the box, and develop creative solutions that align with the company's objectives. It is why they often become high-performing leaders within their organizations.

The combination of technical skills and emotional intelligence makes a powerful combination for engineering success - one that rarely exists naturally in most people. High IQ/EQ engineers are a rarity, but when you find one, you've found a true asset to your team.

How can you spot a high IQ/high EQ engineer?

Well, they're usually the ones who are eager to work through complex problems, take feedback in stride, and can easily switch between technical discussions and conversations about how their actions affect others. They also tend to be confident and open-minded, allowing them to think creatively and develop innovative solutions.

I have a decent track record of spotting these unique individuals with high IQ and EQ using simple discoveries.

How does this person learn?

How do they teach themselves something new? Asking a person to learn something new is a tall order, especially if it's a complex system or technology. Highly effective engineers will have a plan for how they will complete all 4 stages of learning. More succinctly: they know how to teach themselves.

How does this person handle criticism?

How do they respond to a different opinion? It is excellent for interviewing: Ask them a behavioral question – then challenge their outcome. A person without an ideal amount of EQ will get visibly flustered, possibly angry, and be unable to dig themselves out of the perceived mess. You'll know when you see it.

Lastly, learn about how this person handles failure. For example, when talking to a security engineer, I'd ask: "Talk to me about a time you failed in your security career." Get the details, and ask probing questions. Desirable traits in High EQ/IQ people are lack of blame, positive control throughout the scenario, planning, and communication.

High IQ/high EQ engineers are often the difference between a mediocre team and a high-performing one, so it's essential to recognize these skills in potential hires and existing team members. When you have someone with both technical knowledge and emotional intelligence, you have a powerful resource that can take your team to the next level. It is vital to recognize the differences between the two skill sets and facilitate an atmosphere where they can flourish.

Managing high-IQ people

Creating an environment where high-IQ engineers can thrive requires finesse. While it's essential to recognize their technical knowledge and expertise, it's also important to remember that they may need some guidance regarding interpersonal skills and team collaboration.

Leaders should focus on creating an environment where this type of individual can feel comfortable and confident in their abilities. It could include allowing them to take risks without fear of punishment, fostering an environment of open communication, and providing constructive feedback when needed.

High IQ/EQ engineers are often more collaborative and enjoy working with others. It's essential to recognize the value they bring to a team, foster an environment of trust and inclusion, and provide opportunities for collaboration. They also need feedback to grow, so it's crucial to provide it in a supportive and constructive manner.

Who are Low EQ People?

People who struggle with emotional intelligence are those individuals who have difficulty recognizing, understanding, and managing their emotions, as well as those of others. They often struggle in interpersonal relationships because they cannot interpret and respond appropriately to the emotions of others. Low EQ people usually find it difficult to control their impulses. It may make them act impulsively or recklessly in certain situations. They may also have difficulty with empathy and understanding others' perspectives, making it hard for them to form meaningful and lasting relationships. Fortunately, emotional intelligence can be learned and developed over time, so managing low-EQ people is possible.

Managing low EQ people

If you have to manage someone with low emotional intelligence, the key is to understand where they are coming from and be patient. You should explain things more thoroughly and give them extra time to process their emotions. It can also break tasks into smaller steps and provide feedback, so they feel supported in achieving their goals.

Managing people with low EQ, the key is to be mindful of their feelings and respect their emotions. It is essential to provide a safe space to express their feelings without judgment or criticism. Sometimes, you may have to provide gentle guidance regarding certain decisions or tasks that need completion. It's also essential to be an active listener, so you can understand where they are coming from and provide the necessary support. Above all, providing encouragement and recognizing their accomplishments is essential to help them build confidence and boost their self-esteem.

It is essential to mention that all team members, regardless of profession, intelligence, or accomplishment, must be treated and managed as individuals. Use the concept of the emotional quotient as a tool to help you find the best methods to foster successful outcomes with your team.

As a manager, here are some things you should be on the lookout for when managing someone with low EQ;

1. These people need to be sold a story and convinced of the mission:

Some people with weaknesses in this area require clear direction and a compelling narrative behind the tasks they need to complete. It will help them connect with the organization's goals and feel more motivated to complete their tasks.

2. They need support:

Providing constant feedback helps these individuals stay on track, so be sure to take the time to provide encouragement and constructive criticism when necessary.

3. They need to be held accountable:

People can often fall into the habit of procrastination or shirking their responsibilities. You must hold them accountable for their actions and ensure they meet deadlines and complete tasks on time.

4. They tend to be stressed:

They may need to learn how to handle specific tasks or situations, so providing gentle guidance is essential. It could involve breaking tasks into smaller steps, providing resources to help them complete their tasks, or offering moral support during stress.

5. They do not value traditional business acumen and live on their emotional intelligence:

Some people may respond to something other than traditional business tactics, such as aggressive deadlines and strict management. Instead, they are more likely to succeed when provided with a vision directly tied to the value of the effort.

Managers need to recognize this difference and adjust their approach accordingly.

Self-awareness - Knowing your strengths and weaknesses

Self-awareness also deals with knowing your strengths and weaknesses: In the engineering field, this is especially important. When engineers become aware of their weaknesses, they can take steps to address them and learn from their mistakes rather than repeating them. Recognizing emotions in yourself and others is also essential for effective communication and problem-solving in engineering. Engineers can better interact with colleagues, clients, and other stakeholders when they understand and effectively manage their emotions.

For example, if an engineer is feeling overwhelmed by a challenging project, they can take the time to assess their emotional state and reassess what resources or support they may need to get back on track. They need to be conscious of how their feelings affect their ability to think and perform. On the other hand, if an engineer feels confident about their progress on a project, they can use that confidence and optimism to motivate themselves and others to work more effectively.

Understanding the emotions of colleagues and stakeholders is also essential. It helps engineers to be more sensitive and understanding when communicating with others. It could involve the engineer being aware of the subtle nuances in body language and tone of voice facial expressions that can demonstrate what someone else is feeling. Having this emotional awareness allows the engineer to respond in ways that make it easy for them to work together effectively.

The ability to show empathy with others is also essential for engineers. Empathy allows them to recognize and relate to the feelings of their colleagues, clients, and other stakeholders. It can help them build strong relationships and increase trust when collaborating on projects or providing services. By seeing situations from multiple perspectives, engineers can better identify solutions that work for everyone.

Self-regulation - Self-control of actions and emotions

Self-regulation involves an integral part of emotional intelligence. It involves the ability to monitor and manage one's own emotions and reactions to achieve desired goals. In engineering, this means staying calm in difficult or stressful situations and making beneficial decisions for all involved. It could include making sure that deadlines are met or setting realistic expectations for a project.

For example, engineers might feel frustrated with their project progress, but they can use self-regulation to consider the bigger picture. It could involve breaking down the task into smaller steps and working through each part individually. By managing their emotions, they can focus on finding the best solution without letting them get in the way. I like to think of self-regulation in engineering as "thinking before reacting."

It is essential to be able to regulate how emotions are expressed. It could involve refraining from responding with anger or frustration, even under pressure. Instead, engineers should focus on staying composed and professional while communicating with colleagues, clients, and other stakeholders. It can help maintain respect and trust, which is essential for effective collaboration.

Self-regulation also involves being aware of one's impulses and using self-control to act responsibly. It is imperative in engineering since engineers often have to make decisions in high-pressure situations or with limited information. Control of their emotions and actions allows them to make better decisions that benefit everyone involved.

Another aspect of self-regulation is recognizing when one needs help from others. In engineering, this could involve asking colleagues for advice or consulting with clients when making decisions. By being aware of their limitations, engineers can identify areas where they need extra help to achieve their goals. Think of it this way; let's say an engineer is feeling stuck on a project. By recognizing that they need assistance, they can reach out and get the support they need to move forward. Acknowledging your weaknesses and asking for help allows engineers to stay focused on the task.

Motivation - Drive (to succeed)

Have you ever felt motivated to complete a project or task? Motivation is essential in emotional intelligence since it helps drive people toward their goals. In engineering, motivation is vital when completing projects on time and meeting targets. It can also help engineers stay focused and inspired when working on complex problem-solving tasks. It is especially true when working on bleeding-edge technology or projects with ambiguous requirements or outcomes, i.e., Research.

Motivation can come from within or from outside sources. Internal motivators are setting personal or professional goals, staying organized and focused, or feeling accomplished when tasks are completed. External motivators include feedback or recognition from colleagues, managers, or clients.

Regarding engineering projects, motivation frequently is the difference between failure and success. An engineer feeling uninspired or unmotivated will take longer to complete tasks, while someone with a higher level of motivation can stay focused and drive the project forward.

Tapping into internal and external sources of motivation, engineers can create an environment where they are both productive and fulfilled. It could involve setting specific targets, tracking their progress, and rewarding themselves when they meet those goals. Motivation is an essential part of emotional intelligence for engineers, so it's important to take the time to recognize what inspires you and set achievable goals. At this point, I'd like to introduce the "3Rs of Motivation", which stands for Recognize, Remember and Reward. Let's look at each one in more detail.

The first R stands for Recognize: to be aware of what motivates you, whether it is recognition from a colleague or the feeling you get when completing a task. You can start setting achievable goals and creating a plan for success by recognizing what motivates you.

The second R stands for Remember: which is to remind yourself of the benefits of staying motivated. It could remind you why you chose to engineer or a personal goal you want to achieve. It's essential to stay focused on why you're doing what you're doing and constantly remind yourself of the bigger picture.

The third R stands for Reward: rewarding yourself when you complete a task or meet a goal. Recognizing your accomplishments and celebrating the small wins along the way is essential. Whether taking a break, treating yourself to something special, or just patting yourself on the back, rewarding yourself can help keep motivation levels high and create positive reinforcement for achieving your goals.

By understanding and mastering the 3Rs of Motivation, engineers can become more emotionally intelligent and better equipped to tackle challenging tasks. With higher levels of motivation, you'll be able to reach new heights in engineering projects and feel a greater sense of accomplishment. So, take some time today to recognize what motivates you and start making progress on your engineering projects!

Compassion – Understanding

Let's talk about compassion and understanding. Emotional intelligence is not just about being motivated and driven; it's also about having a deeper understanding of yourself and those around you. Compassion is essential to emotional intelligence, as it allows engineers to empathize with colleagues, clients, or even their own mistakes.

Compassion involves taking the time to understand other people's perspectives, as well as understanding your strengths and weaknesses. It also involves self-reflection, looking inward, and assessing how your actions may affect others. Whenever you take time to listen and understand, engineers can develop a greater level of trust with their colleagues and clients, leading to more successful projects. For instance, let's say you just received some feedback on a project. Instead of immediately getting defensive, take a moment to understand the perspective of the person giving you feedback. That way, you can incorporate their suggestions into your work and move forward with greater understanding.

Compassion and understanding are essential skills for engineers to cultivate, allowing us to collaborate more effectively with others. Understanding how your actions affect those around you can help create a more productive working environment and foster healthy relationships. Therefore, take the time to listen and show empathy when needed - it will make all the difference!

Social Skills - Managing relationships of various types

Let's talk about social skills. Social skills are essential for engineers to develop to collaborate effectively with colleagues and manage relationships of various types, whether interacting with clients, negotiating terms with vendors, or simply discussing ideas with co-workers. Having strong social skills is key to success in engineering projects.

At its core, social skills involve effectively managing relationships of various types. It means creating meaningful connections with others and fostering a sense of trust, respect, and understanding. It also involves developing the ability to handle delicate conversations diplomatically and actively listen to those around you without judgment or bias. Let's say you've recently discussed a problem with your boss about a project. Instead of immediately jumping to conclusions, ask yourself what you learned from the conversation and adjust your approach accordingly.

One way we develop social skills is by positively interacting with people. As an engineer, you should communicate effectively with the public and your colleagues to achieve your goals. You can empower yourself by learning to listen and speak articulately, contribute to positive relationships, and help others achieve their goals.

A critical factor to pay attention to is that social skills are cultivated through practice. You can't just sit back and expect to become a master of social interactions overnight. Listening and communicating requires practice, patience, and hard work! Let's look at a few simple ways to develop your social skills.

Communication is a two-way street.

To create meaningful connections, you must take the time to listen and actively engage with others. It involves active listening, effectively hearing what others are saying, and tuning into their emotions and feelings. It also involves speaking clearly and articulately - which is why proper speech training is crucial for engineers! And lastly, it means having the courage to ask questions when needed. Often, the most helpful questions to ask are those that require you to start a conversation rather than finish one.

Relationships are built upon trust.

When building relationships, there are two things to remember: having confidence in what you're saying and being willing to challenge your assumptions. If you speak with someone trustingly, they will find it much easier to listen to you and respond accordingly. And if you're willing to challenge your assumptions, you can avoid making assumptions about others and create a more open mind.

Have self-confidence.

As an engineer, it's easy to think that every suggestion needs to be perfect and every idea needs to be original. It is entirely false - the most successful leaders are those who can step back and listen to the opinions of others. In this process, you can take others' ideas, concepts, and suggestions and make them your own. So when interacting with colleagues, clients, or potential clients, have confidence in your work and stand by it. You never know what little suggestion could lead to a breakthrough! By communicating effectively with others, you will have less difficulty navigating the engineering world and achieve your goals more quickly.

Always be kind.

As engineers, we're often so focused on achieving a goal that we forget to be kind to others. However, kindness is the key to creating meaningful connections with others and setting yourself up for success. Kindness involves looking out for the well-being of those around you and understanding that what may seem like a minor problem could represent a change in perspective or circumstance for someone else. It means being patient, understanding, and forgiving. Engineers are often known as perfectionists - so this last point is essential if you want to become a true leader in the engineering community.

In my experience, strong social skills are differentiators in the engineering market. People who excel at this can often lead on both sides of the business – internally with teams and externally with clients – making them a powerful asset. Some of the most lucrative business deals I've seen signed were the child of an otherwise benign relationship between an engineer and a client executive where trust has been established. If you struggle with this, I implore you to leave your comfort zone and practice being social.

Objectivity - Understanding others

Now let's talk about objectivity. Objectivity is essential for engineers to cultivate as it serves as the basis for developing strong relationships and achieving your goals. Let's start by defining what objectivity means:

Objectivity involves listening to others without judgment or bias, evaluating situations without preconceived assumptions, and communicating clearly and effectively.

Essentially, the role of an engineer is to provide meaningful solutions to issues presented. Therefore, being able to objectively evaluate your surroundings and listen to the problems of others is critical for success. To achieve this, you must be able to take a step back and look at issues from various perspectives. It requires active listening and openness to the situation's facts and circumstances. One way to develop this skill is through learning how to check your judgment at the door when discussing an idea or new concept with others.

Many people are afraid of letting others down by giving them an honest answer - so instead, they opt for a simple yes or no response. However, this is an ineffective way to communicate. Learning how to communicate with others without judgment will help you bring innovative ideas to life and make your work more enjoyable for all parties involved.

To develop objectivity, you must first be able to listen to others clearly and effectively. Active listening involves listening to what others say and tuning into their emotions and feelings without judgment or bias. It also involves being able to ask questions when needed - which can open up entire avenues of conversation that may otherwise have been left untouched. And lastly, it means taking the time to understand other people's viewpoints and make values-based decisions. Through this knowledge, you can create effective solutions for new challenges. Let's look at how this applies to your engineering career.

Develop practical communication skills.

As an engineer, I effectively communicate with others - from management to clients and even co-workers. Speaking clearly and articulately is essential for communicating your ideas and improving your leadership skills. Many engineers rely on technical language to avoid the judgment of others - but this approach could be more effective in communicating ideas!

Be genuinely interested in the work of others.

One of the most significant mistakes engineers make is to assume that everyone else in the organization believes the same way about problems and projects. By genuinely being interested in others' work, you'll be able to build stronger relationships and foster a positive working environment. It will also help you avoid making assumptions about how people feel, allowing you to avoid clashing with others and a better working relationship.

Take time to understand what's important for others.

To not make assumptions, you must take the time to understand what's important for others. Are you on a project with a deadline? Are you on a project that involves more than one person? Do you have other obligations outside of work? By taking the time to understand what's important for other people, you can avoid making assumptions about their feelings and reactions.

Be understanding and forgiving.

While developing objectivity is essential, so is understanding and forgiving in your interactions with others. No one is perfect - everyone makes mistakes! And chances are, people's mistakes result from an honest mistake or oversight. It can be challenging to do when you're frustrated with a project or client - but take some time to step back and evaluate the situation. There's a legitimate reason why things aren't as they should be. By understanding this and taking a step back, you'll be able to set people at ease while still maintaining your integrity. And most importantly, you'll be able to work more effectively as an engineer.

Chapter 6

Career Development

Let's talk about career development. As a manager, it's vital to help your team members develop their skills and career progression. It could involve providing mentorship and guidance on specific tasks, encouraging them to pursue additional training or education, or helping them set career goals and objectives. Additionally, be sure to provide frequent feedback about job performance so that they know exactly where they stand and what they need to improve.

Career development is an integral part of any organization, as it helps team members stay motivated and engaged in their work. The IT industry is constantly changing, so staying up-to-date with the latest trends and technologies is essential to remain competitive. By helping your team members develop their skills and progress in their careers, you are better equipped to create a high-performing team that can keep pace with the ever-changing IT landscape.

Things to keep in mind when developing careers

People always need to feel like they are moving forward (and actually are)

It's vital that people feel like they are making progress and moving forward. It could be through promotions, new assignments, or additional training opportunities. Keeping them engaged and motivated in their work is essential as providing regular feedback and support. Have you heard the saying, "What gets rewarded, gets repeated"? Well, this applies to career development as well.

Invest in employee training

Training is a missed opportunity for many employers, and it's necessary to provide the right tools and resources for someone to grow in their career. It could include additional training courses, access to relevant materials, and guidance from mentors. Providing team members with the right tools will allow them to develop their skills and stay up-to-date on the latest technologies.

People need to be paid fairly and negotiate their wages openly with their employers

When team members are paid fairly for their work, they're more likely to stay loyal and motivated. Don't make a habit of being "penny-wise, pound foolish" and find yourself in a cycle of attrition. Be proactive with reviewing performance and compensation and keep cash compensation in line with the market – or risk losing that all-star to your competitor.

Regularly consider compensation adjustments to ensure that team members are well compensated for their work. It could involve periodic raises, bonuses, or other forms of recognition. Additionally, it is essential to review your compensation practices to ensure they align with the current market and industry standards. Your engineering team is the foundation of your organization, so it's essential to provide them with a fair and equitable compensation package. Here is how it works:

- **Performant + Underpaid team members QUICKLY raised into the median salary range.** It will ensure they receive fair compensation for their work and help motivate them to stay focused on their work and not find new opportunities elsewhere. Investing in your team's success will make them more likely to contribute and stay engaged.

- **Regularly adjust compensation according to market trends** is vital for keeping up with the changing industry standards. It could involve periodic salary reviews or bonuses based on individual performance. Employees who are focused on their payments are not focused on your business – be proactive.

- **Low Performing team members REGULARLY updated on how to achieve their compensation goals, with transparent and fair objectives.** Have you ever heard the saying "what gets rewarded, gets repeated"? Well, this applies to compensation as well. It's essential to provide low-performing team members with transparent and fair objectives they can use to reach their goals. It means regularly updating them on their progress while giving them the necessary resources to succeed. In a situation like this, it's crucial to provide team members with an opportunity to grow and develop.

Setting goals that tie to compensation is a big challenge for everyone involved – the business that needs justification, the manager who is unhappy with the performance, and the employee who feels underappreciated. A people-first approach, which is a win for everyone, is to provide specific, measurable, achievable, relevant, and timely (SMART) goals. You want them to succeed, make it happen.

All satisfactory performing team members are given opportunities to earn additional money.

While providing fair and equitable compensation for your team, it's also essential to allow them to add additional value to the business. It could involve providing team members with opportunities to achieve stretch goals for a project or client that directly ties to additional business revenue. Encourage, without burning out, your team members to push the boundaries of success and earn a few extra bucks. By allowing satisfactory performing team members to earn extra money, you can motivate them to stay engaged and remain loyal to your team. When team members are allowed to increase their income, they're more likely to stay motivated and contribute to your business's success.

An exciting way I've seen this model succeed is with certifications. Many business key performance indicators are tied to certification or compliance objectives. For example, Amazon Web Services partners require a certain number of engineers on staff to maintain certifications in specific areas. Many companies encourage team members to achieve these certifications (and keep them current) by tying them to additional compensation. For example, a $1,000 bonus for every professional level certification earned that can be registered to the company.

High-performing team members praised and recognized

Finally, it's essential to praise and recognize your high-performing team members. It could involve public recognition or other forms of reward. By letting your team know that their hard work is appreciated, you can motivate them to stay loyal and contribute to the success of your organization. When people are acknowledged for their hard work, they are more likely to stay engaged and continue to strive for success.

Giving praise can feel awkward, and it isn't easy to achieve with perceived sincerity. To overcome this challenge, try using a product demo or project status update to give public praise and encourage others to participate.

How bonuses and rewards programs work

Bonuses and rewards programs are a great way to motivate your team and give them the recognition they deserve. It's essential to set clear goals for team members so that they can understand what is expected of them. Here are tips to help you implement a successful bonus and rewards program:

Employee referrals

Encouraging team members to refer their friends and colleagues can be a great way to reward team members for their hard work. It could involve providing bonuses or other forms of recognition for successful referrals.

Cost savings

Providing team members with incentives for cost savings can motivate them to stay engaged and look for creative solutions.

Rewarding team members for finding ways to save money will also show them that you value their contributions.

Revenue generation (e.g., sales influence)

Revenue generation is another excellent way to reward team members for their hard work. It could involve providing bonuses or other forms of recognition for successful sales or other revenue-generating activities. For instance, awarding overachieving engineers who put their all into closing a deal with a small portion of the commission. It will help to motivate the team and acknowledge their hard work.

Team performance

Rewarding teams for successful projects or campaigns can also be a great way to motivate team members. It could involve bonuses or other forms of recognition based on overall team performance. What's important is that everyone gets recognized and rewarded for their contributions.

Incentives

Equally important is to provide incentives for team members to stay engaged and motivated. It could involve providing gift cards or other rewards when they reach certain milestones or complete specific tasks. By providing incentives, you can ensure that your team members continually strive for greatness and stay loyal to the organization. Here are a few ideas for incentives you can use:

Formal Training

Provide team members with the opportunity to attend workshops, seminars, or other formal training programs. It could involve offering discounts on courses or providing free access to online learning materials.

Company swag

Recognize and reward team members with company swag such as t-shirts, hats, or other branded items. These items can create a sense of pride and loyalty within the team.

Flexible work arrangements

Providing flexible hours or other accommodations can be an excellent way to reward team members and keep them motivated. It could involve allowing telecommuting, paid time off, or other arrangements that make it easy for team members to balance work and family life.

Social events

Organizing social events and team-building activities can be a great way to reward team members and show your appreciation for their hard work. It could involve dinners, movie nights, or other enjoyable outings that will create a sense of camaraderie among the team.

Recognition

Simply saying thank you for a job well done will go a long way in motivating your team. You could also publicly recognize team members through a company newsletter or other channels. It will show that you value their contributions and are grateful for their hard work.

Public speaking

Allowing team members to present at conferences or other public speaking events greatly rewards them and provides recognition. It will also help build their confidence and create an atmosphere of mutual respect among your team.

Marketing (e.g., company-sponsored events, podcasts, training, travel, etc.):

Providing team members with the opportunity to attend company-sponsored events, participate in podcasts, receive training, or even travel can be great for motivating them and building morale. It will show that you value their contributions and are willing to invest in them. Have you ever considered providing these types of incentives?

Overall, the goal of implementing rewards and incentives is to create a culture that motivates team members to stay loyal to the organization. By providing meaningful rewards, you can ensure that your team feels valued and appreciated for their hard work.

It's important to understand that it takes more than money to motivate an employee. It's also essential to create a culture of recognition and appreciation that encourages team members to stay loyal to your organization. By providing meaningful rewards, incentives, and recognition, you can ensure that your team is motivated and working hard for your business.

Pro Tip: Encourage team members to speak up with any questions or concerns about salary directly to you rather than discussing it with each other. It will help ensure that salary conversation remain collaborative and free from adversarial elements. Please remember not to publish individual employee salaries as this is confidential company information.

How the engineering team can impact a company's bottom line

An engineering team can directly impact a company's bottom line. They are responsible for developing and maintaining the technology that drives the business forward. Whether creating new products or services, improving existing ones, or helping optimize operational processes, their work plays a crucial role in increasing company profits and growth. Here are a few ways engineers can help drive a business's success:

Influence

Give them the opportunities to flex their intelligence with executives or customers:

If there is anything engineers are known for, it is their intelligence. They bring a lot of knowledge and problem-solving capabilities to the table that can help drive business success. Allowing engineers to engage executives or customers in conversation will allow them to show off their expertise and influence decisions. It can also help build trust between the engineering team and the business. Here is a scenario to help you understand:

Your engineering team has worked tirelessly to develop a new product. When the engineering team presents it to the executive team, the engineering team can help influence the executive team's decision by using data and explaining how their solution can solve the current problem. It will demonstrate their knowledge and build trust between them and the business.

Allowing them to influence policies or business decisions

Engineers have a unique perspective that can add value to any business. Allowing them to influence policies or business decisions will help ensure that the engineering team is heard and respected. Additionally, this will create an environment of collaboration between engineers and the rest of the organization. It can also lead to improved morale within the engineering team as they feel included and valued. Here is an example to consider:

Your engineering team has identified a bug in the system that affects customers. Allowing them to present their findings to the executive team and suggest changes or improvements will show that you value their insight and contributions. It can create an environment of trust between the engineering team and the business, leading to improved morale and loyalty.

Encourage the formation of committees or guilds focused on business topics

Business topics should include more than just the executive team. Encouraging your engineering team to form committees or guilds focused on specific topics such as customer experience, product development, and quality assurance will help foster collaboration between them and the business. I like to think of these guilds as the "brain trust" of your organization. They can bring valuable insights and knowledge to crucial business decisions, helping drive success for the engineering team and the whole company.

Chapter 7

Hands-On Leadership

"IF YOU THINK YOU'RE LEADING, BUT NO ONE IS FOLLOWING, THEN YOU ARE ONLY TAKING A WALK." - JOHN MAXWELL.

Hands-on leadership is a style of management that puts the leader near the team. It allows the team leader to see what's happening at all levels and act when issues arise. Being readily available will allow you to have quicker responses and be more proactive in handling problems. Hands-on leaders are also known to be "emotional" leaders, which means they can take responsibility for their team's mistakes. As a hands-on leader, you should be prepared to bear the consequences of your team's actions by getting involved in the solution.

These are the few things you will need to do to become a hands-on leader:

Don't micromanage

Micromanaging is not an effective form of leadership. Sometimes we may be tempted to micromanage because we want everything to go smoothly. But in reality, this can hinder progress and innovation because it takes more time and effort than conventional methods. When we micromanage, we are not allowing our team members to grow and become independent. Instead, we spend too much time working in the same areas.

Being "Hands-On" is beyond coding and crisis management

Being "hands-on" is more than slinging code or sitting in the war room at crunch time. It's about being engaged in business, engineering, and with the customers/end users. It's about being proactive and accountable and taking an interest in your teammates, not just their code. It is a more profound, lasting, and meaningful relationship. Giving attention to these areas will help your team function smoothly and efficiently.

Don't Command from an Ivory Tower

Have you ever seen a leader who stays out of sight but gives orders from an ivory tower? It might work in the movies, but it won't work in engineering teams. Commanding from an ivory tower will make your team feel isolated and disconnected. It will also hurt their morale because they will not be able to express their ideas or provide feedback without fear. People will do their work, but there will be better work that they can offer.

Furthermore, you will lose your technological edge and the trust of your team if you don't stay involved with them. For instance, let's say your team is working on a challenging project requiring technical know-how and collaboration. If you are not actively involved in the process, your team will struggle because they won't respect your input or guidance. You might even find a few innovations only you can bring to the table!

Reacting Negatively, Displaying Bias, or Lacking Context

Also, your team members will cease to bring problems to you or discuss them openly if they know that you will react negatively or without considering the full context of the situation. Instead, it's essential to be open and willing to listen to their perspectives before making any decisions. Doing this will foster trust and loyalty among your team members and help create an environment where they can safely express themselves without fear of being judged. What is a manager without the trust and respect of their team? Nothing.

Embrace Change

Change can often be scary, especially for engineering teams. As a leader, you must embrace change instead of trying to resist it. It means having an open mind when new ideas are brought to the table and being willing to adapt to changes in the industry. Additionally, it means providing feedback and constructive criticism when needed and embracing a culture of continuous improvement. It is equally important to be flexible in your management style and adjust your approach as needed. It will show your team that you are open to new ways of doing things, which can help foster growth and innovation.

Engineering teams need to be at the front of a fast-changing world. Embracing change and staying open-minded can help your team become more agile and adaptive, leading to better results in the long term. So don't be afraid to shake things up a bit!

Lead by Example

Influential leaders are the ones that lead by example. As a leader, you should always strive to set the standard for your team regarding quality and professionalism. If your team sees that you are motivated and passionate about your work, they will be likely to put their best foot forward.

It is challenging to quickly learn and implement, as you, as a leader, need to be motivated and respected in your role. Consider sending a copy of this book to your boss!

You must set aside some time to be hands-on

Leading an engineering team doesn't mean you have to be a technical expert in every field. Instead, it means allowing yourself to get your hands dirty and stay abreast of the latest trends in technology. It includes taking time each day or week to check on progress and provide input, engaging with your team members, and offering advice and feedback when needed.

Being active in the process will demonstrate to your team members that you are committed to their success and may help you spot potential problems earlier. After all, there is nothing more valuable than having an experienced leader with a holistic view of the project who can quickly identify areas of improvement. You don't need hours of your day; even two hours dedicated to getting involved with different teams can significantly impact you.

You can be engaged in various ways

Write code: Even if it's to brush up on your coding skills, solve a problem that you know is related to the project, or refactor a piece of code that needs improving. Doing so will demonstrate to your team that you are engaged and committed, and you can also take the load off a team member.

Liaise: Translate customer needs and engineering needs between the two teams so both sides are better informed and understand each other's requirements. Doing this will help ensure all involved are on the same page and working together towards a common goal.

Develop: Develop and refine requirements, and build a working knowledge of what's being engineered and how it works. Doing this will not only help you understand your team better, but it will also give you a much better understanding of the challenges they are facing.

Exhibit: Show your team you are present and engaged by attending events, conferences, hackathons, and more. Doing this will make your team feel valued, and they will be more likely to respect your opinion since you are experienced in the industry.

Leading an engineering team with purpose is no easy task, but with a purposeful approach and the right strategies, you can create an environment that encourages collaboration and innovation.

Why should you implement hands-on leadership?

Now, why should you put in the effort to implement hands-on leadership? The answer is simple: it allows you to create an environment where everyone feels heard and respected, which can help build a culture of trust. When team members feel appreciated and involved in decision-making, they are more likely to be engaged and motivated.

In addition, hands-on leadership will help strengthen the relationship between you and your team. By getting to know your team member's strengths and weaknesses better, you can ensure that everyone is working towards the same goal. It will boost morale and productivity and give your team a sense of accomplishment and pride in their work–which often leads to more innovative solutions.

From my experience, some of the best organizations to work for and those who unironically is the most profitable have hands-on leadership. It means everyone is an individual contributor to some degree.

When hands-on leadership is not the best option

Although hands-on leadership is recommended for most situations, it could be better in some situations. An example is if your team is working on a project requiring a lot of collaboration and brain power, let your team lead the charge, allowing them to experiment and explore new ideas. If you are too involved with the project, it may stifle creativity and innovation. Remember that hands-on leadership doesn't mean micromanaging or being overbearing but instead offering support and guidance when needed.

I like to think that hands-on leadership is like a bridge—it can help you connect the different parts of your team and ensure everyone is on the same page.

Chapter 8

The Future of Leadership

It's time we redefine leadership. It doesn't have to be a top-down approach anymore; it can be a collaborative effort between managers and engineers. We must promote an environment that encourages communication, understanding, and trust to foster great engineering teams. As technology evolves, so should our leadership styles - we now have the power to shape our engineering cultures into something that promotes creativity and collaboration.

The future of leadership is in the hands of those willing to take a risk and make changes. When the right tools and strategies are adopted, you can create teams that trust each other, look out for one another, learn from their mistakes, and grow together as engineers. You need to understand your strengths and weaknesses to address your teams' different needs better. It would help if you also strived to create an atmosphere that encourages communication, understanding, and trust between managers and engineers.

The engineering leadership challenge

The engineering leadership challenge is a common issue faced by many teams. Without formal training and development, engineers are often thrust into management roles without the tools necessary to excel in their new responsibilities. It may seem logical for the best engineer on a team to become the leader. However, being a practical engineer only sometimes translates into being an effective project manager or people manager.

This mismatch of skills can lead to poor performance from the new leader, frustration from the team members, and, ultimately, burnout. The leader and their team suffer as they grapple with how to foster leadership within the engineering organization successfully. Organizations need to provide training and support for engineers transitioning into leadership roles. It allows them to learn the necessary skills and develop an ongoing support system. Think of it this way: the better-prepared leaders are, the higher their chances of success.

The Catch

The catch is that while it can be daunting to transition into a leadership role, becoming an effective leader provides excellent rewards. Not only does this help engineers become better leaders and managers, but the team can benefit too! With the proper training, mentorship, and guidance, engineering teams can foster a culture of open communication and collaboration where everyone feels empowered and capable of achieving success.

Engineering teams need support to ensure their members are prepared for smooth transitions when taking on management roles. It requires organizations to invest in formal training programs and ongoing mentorship and development opportunities so that engineers feel equipped to lead their teams effectively. It's an investment worth making, one with potentially huge returns in terms of team performance and morale.

Training a manager is very expensive, especially from the ground up. I often face heavy pushback on training budgets for managers as it's just another dollar in the cost center for the leadership team. However, the cost of having a poorly managed engineering team is fatally higher – staff attrition and the failure of the business are the stakes.

By investing in the development of their engineers, organizations can create a culture where teams are empowered to take on challenges with confidence while fostering solid relationships based on trust and collaboration. Ultimately, this helps engineering teams achieve their objectives more efficiently and effectively — everyone wins!

Chapter 9

Lessons From 8 Successful

Engineering Cultures

The modern workplace is evolving rapidly. Companies that want to maximize their potential must look for ways to make the most of this new environment. Dan Pink said, and I quote "People are more productive and fulfilled when they have autonomy, mastery, and purpose." As such, successful high-tech companies have published documents detailing their company culture to inspire others and share what is working well for them. Examining these documents can give insight into the best practices of modern workplaces in order to get the most out of the new work environment. Autonomy, mastery, and purpose are the three key elements that lead to success in the world of work.

The company cultures highlighted in this chapter reveal common themes and values that, while not feasible for everyone to implement, can help inform and guide other organizations' approaches to management and productivity. I have carefully selected some of the world's most successful high-tech companies to provide various examples and insights applicable across industries.

The critical elements for success that these companies have in common include: fostering creativity, promoting collaboration, embracing diversity, encouraging feedback and open communication, creating meaningful work through goal setting and clear expectations, offering competitive benefits, and providing a sense of autonomy to team members. These elements are the foundation upon which these successful companies have built their cultures and can provide valuable guidance for any organization looking to transform its culture. Let's explore these in more detail.

Creativity is critical to successful company cultures, allowing team members to generate new ideas and explore innovative solutions. To foster creativity, companies provide platforms for collaboration, such as brainstorming sessions or hackathons. They also allow for experimentation and risk-taking, providing team members the space and support necessary to explore new approaches.

Collaboration is another crucial element of successful software engineering cultures. Companies that promote collaboration ensure that they create a sense of community and provide teams with the resources needed to support each other in their work. They also emphasize communication between all levels of staff so that ideas can flow freely and team members know what to expect from each other.

Diversity is another vital component of successful software engineering cultures. By ensuring that all voices are heard, and everyone feels valued for their unique contributions, companies can create a positive work environment where everyone is respected and exchanges ideas freely.

Feedback is also essential to successful company cultures, as it allows for continuous improvement and adaptation to changes in the market. Companies should encourage team members to provide feedback on their work and how teams are performing as a whole.

Finally, successful software engineering cultures offer autonomy and empowerment to team members through goal-setting and clear expectations. By establishing clear goals, companies can ensure that everyone is working towards the same objectives. When paired with regular feedback and rewards, goal setting can create an environment where team members feel empowered to take the initiative and drive progress. Below are the seven companies and their innovative approaches to software engineering culture.

1. *Netflix*

What does your mind picture when you think of Netflix? The streaming giant has revolutionized the way we watch our favorite movies and shows. But what's truly remarkable is how they have created a culture of innovation and collaboration among their engineers. At Netflix, team members are encouraged to take ownership of their work while receiving feedback from their peers to refine their work continuously.

At the heart of Netflix's culture and business strategy lies a belief in freedom coupled with responsibility. They strive only to hire those accountable enough to handle this freedom. This approach has enabled them to stay competitive and agile despite the constantly changing landscape of the software industry.

Netflix is also the pioneer of microservices, a concept that has revolutionized how software is built. Every developer at Netflix is responsible for writing their tools, fixing them if they break, and creating the necessary documentation and operational pain points. It encourages developers to claim ownership of their work and learn from any mistakes that may arise during the process.

Building successful software engineering cultures, Netflix serves as a great example. By placing trust in its team members and giving them the freedom to innovate without bureaucracy, the company can stay ahead of the competition and continue creating great products for its customers.

Let's look closely at some of the unique things we can learn from Netflix's approach to software engineering.

132

- Outstanding software engineering cultures rely on freedom and responsibility.
- Developers should take ownership of their work and learn from mistakes.
- Companies should trust their team members to innovate without bureaucracy.
- Competitiveness and agility are essential in a constantly changing landscape.
- Microservices have revolutionized software development.
- Companies should strive to create great products for their customers.

Let's delve into each of these lessons to understand how you can use them to create successful software engineering cultures in your company.

Freedom and responsibility: You can attract the best talent by granting your team members the freedom to be creative and flexible in their work. However, only those who are responsible enough must be given this freedom. That way, you can ensure that you have a workforce capable of handling the growth and changes necessary for your business to thrive.

Ownership: Developers should take ownership of their work, from writing the tools to fixing any issues that may arise. It proves beneficial as it gives them a better understanding of their product and encourages them to learn from their mistakes.

Trust: Have you heard the phrase "trust but verify"? Companies should strive to trust their team members to innovate and create incredible products while still staying aware of the progress being made. It helps ensure that your business stays ahead of the competition while giving team members the freedom to do their best work.

Competitiveness and agility: Companies need to be competitive and agile in the ever-changing landscape of software engineering. By giving your staff the freedom to innovate without bureaucratic restraints, you can ensure that your company is always one step ahead of the competition.

Microservices: Netflix has revolutionized how software is built by pioneering the concept of microservices. Each developer is responsible for writing their tools and fixing any issues that may arise. It encourages developers to claim ownership of their work and learn from mistakes they might make.

Great products: Ultimately, companies should strive to create great products for their customers. You can ensure that your customers are getting the best possible service through creative problem-solving and a focus on customer satisfaction.

2. *GitLab*

GitLab's focus on comprehensive tools and pipelines for every stage of software development is mirrored in its employee handbook. The handbook clearly outlines GitLab culture and values and offers detailed steps to onboarding and workflows for developers, marketing, sales, finance, and more.

GitLab's engineering teams are typically small, consisting of four developers covering front- and back-end development and UX and product management. GitLab encourages its developers to work even if they specialize in one area. A miniboss or end boss will check their work and approve code merges.

GitLab takes a remote-first approach to its operations, meaning that all workflows are constructed around remote work as the default, regardless of whether team members are co-located. Employees have ample freedom to choose the tasks they work on and how they complete them. The company is committed to giving its team members plenty of space to make decisions and be creative.

Themes: GitLab encourages its engineers to build the tool they use and become domain experts. It is why "dogfooding" is part of GitLab's culture. They even encourage team members to submit merge requests for changes to their online handbook, just as one would with an open-source GitHub project. Below are some themes from their handbook.

Including diversity and inclusiveness:

GitLab understands that team members have different specialties and skill levels. Employees should feel comfortable admitting when they don't understand something or lack an answer without feeling embarrassed or demeaned by one another. They also encourage honest and direct communication, even if it involves job dissatisfaction.

GitLab only cares about output and results, regardless of how much time is spent on the work or if team members check Facebook and Twitter throughout the day.

Transparency is also vital - everything GitLab does is public by default, just like their handbook. It allows for collaboration and innovation.

GitLab encourages using simple, boring solutions as they are often the minor complex, easiest to maintain, and most efficient approach. Iteration, innovation, and complexity can always be added if necessary.

GitLab promotes kindness, empathy, and respect in the workplace. They encourage team members to be kind and direct but still respectful of one another. GitLab hopes to cultivate an innovative culture that values output and growth by fostering an environment of support and collaboration.

Lastly, GitLab encourages "quirkiness," which makes the work environment more exciting and celebrates diversity and different personalities. Open-source projects are a community, so they want to encourage a community of open and giving people.

3. Facebook

The "little red book" by Facebook, created in 2012 when the company passed one billion users, is a handbook designed to explain the company's mission, history, and culture to new team members. It was aptly named after the historical "little red book" written by Mao Zedong that contained speeches, quotes, and writings embodying his philosophy. Similarly, Facebook's little red book contains pithy stories and aphorisms that illustrate the company's values and mission. Unfortunately, only a few pages of this handbook have been made public. Still, it is an excellent example of how successful software engineering cultures document their culture to ensure that all stakeholders understand the mission and values of the company.

This document serves as a reminder to team members of the culture they are joining while also providing a helpful reference for people who want to learn more about Facebook's unique approach to software engineering. Moreover, it is yet another example of how valuable having a documented culture can be for any organization serious about success.

Given the success of Facebook, it's no wonder that many other software engineering companies have followed suit in documenting their culture. Doing so ensures that all stakeholders know the organization's mission and values, which can only lead to better results for everyone involved.

The stories included in Facebook's little red book are meant to inspire new team members and show them how simple ideas can change the world. It states that by changing communication, Facebook is indeed changing the world - a powerful reminder of the company's mission. This sentiment is backed up by an impressive statistic showing that they have approximately one engineer for every one million users. Such a statistic not only highlights the impact of each engineer but also emphasizes how vital each team member is in making Facebook so successful. It helps new hires understand that their contributions are valued and appreciated by the company.

Themes: Facebook's little red book emphasizes the need to stay ahead of competitors and stresses that "the quick shall inherit the earth." Management believes in releasing fast, failing fast, and learning fast to stay caught up. It is why they don't have a long-term five-year plan but a six-month plan and a constantly shifting thirty-year plan. It helps ensure that they remain agile and can adjust quickly to changes in the industry or customer needs.

In other words, such a flexible strategy has undoubtedly helped Facebook stay ahead of its competition and become one of the most successful software engineering companies.

The passages in Facebook's little red book emphasize a healthy paranoia and a sense of tough love. They want to stay ahead of the competition by reminding themselves that "if we don't create the thing that kills Facebook, someone else will." The book also recognizes that there is often a trade-off between greatness and comfort - similar to the tone in Netflix's cultural slideshow. Greatness is the goal at Facebook because they make money to build better services, not to make money. This sentiment is summed up perfectly by the quote: "Fast doesn't just win the race. It gets a head start for the next one". It reinforces the idea that agility and responsiveness are essential if Facebook wants to remain ahead of the competition.

The sentiment here is that each team member is valued and appreciated, which encourages new hires to contribute to the success of Facebook.

Overall, Facebook's little red book provides a great insight into how the company approaches software engineering and culture. It emphasizes a mentality of remaining agile, staying ahead of the competition, and valuing each team member's contribution. These lessons can be applied to any thriving software engineering culture and provide an important reminder of what it takes to remain at the zenith of a rapidly changing industry.

4. Buffer

Buffer is known for its highly transparent company culture. From the CEO's salary to their team's decisions and experiments, team members are encouraged to blog about everything inside Buffer. This level of transparency sets them apart from other companies. For example, anyone can find out exactly how much any member of the Buffer team makes - in 2016, the CEO's salary was $218,000. In addition to salary transparency, the small company of 60+ people is constantly experimenting and making decisions with their team in full view of everyone who works there. This culture has helped Buffer become one of the most successful software engineering companies.

This level of transparency builds trust between team members and encourages creativity and collaboration. It helps team members feel inspired to think outside the box and come up with new ideas that can benefit the company. With this kind of environment, it's no wonder Buffer was able to develop its social media post-scheduling tool into a popular tool in the market today.

Themes: Buffer is all about transparency. They believe in being open and honest with their team members, which is why they publicly share information like codes, revenues, diversity stats, employee salaries, and even how they calculate salaries. This level of openness helps to build trust between team members, as well as encourages collaborative thinking. With this kind of culture, it's no surprise that Buffer was able to develop its social media post-scheduling tool into one of the most popular tools on the market today. Transparency also allows everyone in the company to understand precisely how decisions are made and why creating a clear understanding of the company's direction. This kind of culture has been integral to Buffer's success as one of the most successful software engineering companies.

Buffer's commitment to transparency extends to its salary policies. Salaries vary based on experience, role, and cost of living, but they also offer team members a 5% raise every year of service. On top of that, Buffer has a comprehensive list of all their employee's salaries (first names only) available online, which include roles, start dates, location, and salaries. They also value the feedback from their team members, which is why when negative feedback got to the management about one policy that gave team members an extra $3,000 per year for each dependent, they changed the formula to move that money into grants and other family support benefits.

Diversity is another area where Buffer puts in a lot of effort. They have created a diversity dashboard, an incredibly data-rich visualization that tracks team members and applicants regarding gender, race, and other factors. It allows them to effectively measure their progress towards closing the women and minorities gap in tech. Buffer believes that if you measure diversity, it will be addressed and improved - which has proven true in their case. Through this commitment to diversity and transparency, Buffer has created an environment that encourages collaboration and innovation - making it the highly successful software engineering culture it is today.

5. *Basecamp*

Take a closer look at Basecamp—a project management software created to help teams "get organized and stay on top of their projects."

Basecamp's core values are customer obsession, creative problem-solving, delighting customers, and having fun while getting the job done. These values have been fundamental in creating an engineering culture where people are motivated to work together and challenge each other.

Basecamp, formerly known as 37signals, is widely renowned for its software engineering culture. It develops web-based project management and customer relationship management (CRM) tools that many companies have used over the years. Its most famous team members are David Heinemeier Hansson (aka DHH), the creator of Ruby on Rails, and Jason Fried, who has co-authored two influential books on business (Rework) and remote working (Remote: Office not Required).

The lessons learned from Basecamp's highly successful software engineering culture can serve as an excellent source for other companies to consider. They include understanding the importance of collaboration between team members, having clear communication channels, and creating an environment that encourages creativity. Additionally, Basecamp values the individual contributions of each team member through recognition and rewards. These elements are essential for companies to consider when looking to create a thriving software culture.

Themes: Basecamp is an example of a highly successful software engineering culture, with several core values that make its working environment unique. Like how Netflix attracts talent with generous salaries in the top 5% of the software industry, Basecamp also rewards its team members for their hard work. In addition to excellence, experimentation, honesty, and kindness, Basecamp also has a culture of charity.

Management encourages team members to use their skills and knowledge to improve the world by donating their time and money. This commitment to giving back makes Basecamp stand out from other software engineering cultures. It highlights the importance of individual contributions beyond monetary rewards. Additionally, team members are given ample recognition and rewards for their hard work. These elements are essential to building a thriving software engineering culture.

With its focus on keeping team members healthy, happy and well-rested, Basecamp institutes "summer hours" from May 1 to August 31. During this period, most team members work Monday through Thursday only. The company also provides generous financial support for its staff in the form of a $100/month fitness stipend and a $100/month massage allowance. Also, it includes a community-supported agriculture allowance. These benefits are designed to improve the well-being of team members but also to create a work environment where team members feel safe and appreciated for their hard work.

Additionally, Basecamp is committed to its team members' physical and mental health by providing these additional incentives. Companies must recognize that their team members' health and happiness are essential for a thriving software engineering culture. These incentives demonstrate the commitment of the company to employee satisfaction.

"WE DON'T WANT PEOPLE WORKING MORE THAN 40 HOURS A WEEK IN ANY SUSTAINED FASHION (WE EVEN BUILT IN A 'WORK CAN WAIT' FEATURE IN BASECAMP 3, WHICH TURNS BASECAMP NOTIFICATIONS OFF AFTER WORK HOURS AND ON WEEKENDS). IN A CRISIS, OR A ONCE-EVERY-COUPLE-YEARS SPECIAL PUSH, WE MAY REQUIRE VERY SHORT-TERM EXTENDED HOURS, BUT OTHERWISE, WE STRONGLY ENCOURAGE A MAXIMUM OF 40 HOURS A WEEK, AND 8 HOURS OF SLEEP A NIGHT." —JASON FRIED.

According to Jason Fried, co-founder of Basecamp, the company does not encourage its team members to work more than 40 hours a week. This commitment to giving team members adequate time for relaxation shows that Basecamp understands the importance of having a healthy work/life balance. Furthermore, they have instituted a 'Work Can Wait' feature in their software where notifications are turned off after work hours and on weekends. This feature allows team members to enjoy their time away from the office without having their work follow them home. Basecamp is committed to its team members' mental health, essential for a thriving software engineering culture.

Basecamp has intentionally kept itself a small company, with less than 50 team members, as it believes this reduces complexity and increases efficiency. The company is fiscally conservative and has remained profitable for 15 years in a row. Co-founder Jason Fried also encourages a work week of just 40 hours; he strongly opposes the industry standard of asking team members to work more than 60 hours a week and regular weekend shifts. It speaks to Basecamp's commitment to ensuring that its team members have some life outside their work. The company also has a culture document called "Culture crushin' on Basecamp," which outlines the values and culture of the company. This document emphasizes why Basecamp is a thriving software engineering culture that has achieved such success for over 15 years.

6. Etsy

Etsy, the online arts and crafts marketplace, has a service-oriented architecture often mistakenly referred to as monolithic. Its engineering culture emphasizes a disciplined continuous delivery process with a rapid feedback loop of about 21 minutes (which may be even shorter now). It allows developers to deploy code 25 times daily, providing an invaluable sense of accomplishment for everyone involved. By recognizing and rewarding its engineers' hard work, Etsy has proved to be a highly successful software engineering culture.

A crucial part of Etsy's success is its focus on providing a great development experience - something that other software engineering cultures can benefit from learning. They emphasize automated testing, continuous integration and delivery, and effective collaboration practices. It helps developers to have more control over their work and ensures that the result is of high quality. Moreover, by constantly looking for ways to improve its processes and systems, Etsy has consistently delivered excellent products while keeping its engineering teams engaged and motivated.

Themes: Etsy believes that engineering processes that make developers happy are also the best for product quality. To this end, they have implemented many practices to ensure their developers feel empowered and impactful from day one. For example, new engineers can add themselves to the team page on their first day - which teaches them about the deployment process and gives them a sense of accomplishment. Moreover, if their code introduces any bugs, the company has a policy of blameless post-mortems to ensure that developers are not held accountable for failures. It helps keep morale high and encourages engineers to take risks without fear of negative consequences. Overall, these practices have made Etsy a highly successful software engineering culture.

Regarding software engineering, Etsy is proving that success only sometimes requires the latest and greatest technology. Instead, they prioritize processes and culture over flashy tools—and it's working! By focusing on providing their developers with a great experience through automated testing, continuous integration and delivery, effective collaboration practices, and more - they've been able to keep their engineers engaged and motivated. No wonder they are setting a well-deserved example for modern software engineering cultures everywhere!

Etsy emphasizes the importance of being able to 'feel' the impact of their work every day, citing Dan Ariely's book, The Upside of Irrationality: The Unexpected Benefits of Defying Logic at Work and Home. It is especially true for companies that practice continuous delivery - where new code can be deployed quickly and easily, with minimal disruption. It allows developers to feel the tangible results of their work and motivates them to continue striving for excellence.

"If companies want their workers to produce, they should try to impart a sense of meaning—not just through vision statements but by allowing team members to feel a sense of completion and ensuring that a job well done is acknowledged. Such factors can hugely influence satisfaction and productivity."

The statement above perfectly summarizes Etsy's highly successful software engineering culture. By allowing developers to feel their impact and recognize accomplishments, they can keep morale high and continue innovating. This type of environment can help any company achieve optimal productivity and satisfaction in their software engineers.

Like other successful software engineering cultures, Etsy's organizational model is flat. It creates a sense of autonomy and responsibility amongst developers and operations teams, who are encouraged to collaborate and watch out for each other. With a "radical decentralization of authority" in place, the company trusts its team to move projects forward - giving them the freedom to make decisions and take risks. It allows for more significant creative thinking and innovation, leading to more impactful results.

Additionally, its self-regulating nature fosters a sense of community within the team - making it easier to stay motivated because they know they are part of something bigger than themselves. Ultimately, this type of culture has been proven to be highly successful and can be replicated in any software engineering organization.

Empathy and diversity are core values at Etsy, with the company doing its part to promote female engineers in tech. It includes sponsoring a summer hacker school that offers grants to women - a move that saw an impressive 500% increase in the number of female engineers employed by Etsy in 2014. Furthermore, they also offer team members exiting the company the opportunity to give a post-mortem presentation - a chance for them to share their knowledge, experiences, and insights they gained during their time there. This practice is unique amongst software engineering organizations and serves as an example of Etsy's commitment to cultivating a positive culture where developers feel valued and appreciated.

As co-founder Robert Neumeier famously said, "Optimize for developer happiness" - a mantra that other software engineering organizations should consider.

7. *Zaarly*

Zaarly offers the perfect example of a software engineering culture that values flexibility and remote work. Unlike most companies, they don't have strict vacation policies and instead allow their team members to set their schedules without adhering to a predefined set of rules. Additionally, if in-person meetings are required, Zaarly is more than willing to cover the cost of plane tickets. Based in Kansas City, Zaarly offers its team members the freedom and flexibility to create a work/life balance that works best for them. Their culture puts team members' needs first, making it one of many successful software engineering cultures.

At Zaarly, there is no formalized hierarchy, and team members are encouraged to reach out and collaborate regardless of their experience level. Everyone is free to work whenever and wherever they feel most productive. Employees can set up meetings with anyone on their first day if needed - because everyone's voice matters in this highly successful software engineering culture. This open and collaborative environment is key to their success.

Theme: With communication being a core focus at Zaarly, the handbook outlines the advantages and disadvantages of every form of communication, from email to face-time. While remote work is encouraged, full collaboration can only happen when people are in the same room. Zaarly understands the importance of good communication, as they have noted that 9 out of 10 internal conflicts arise due to miscommunication or a lack thereof. This emphasis on communication is integral in creating a thriving software engineering culture and achieving success. Ensuring everyone is on the same page allows the team to work more efficiently and effectively with little room for misunderstandings or conflict.

Prioritization is an essential priority for Zaarly. Their management understands it takes time to identify the right priorities and encourages feedback from half of their team if needed. They believe in employing data or customer feedback to settle arguments about priorities or other issues whenever possible. However, when definitive customer feedback or data isn't available, Zaarly encourages team members to trust their gut. They believe that even though mistakes could be made, following one's intuition can lead to some great insights and valuable lessons that will help further the success of their software engineering culture. By relying on factual data, customer feedback, and intuition at different levels, Zaarly can determine its priorities and make the right decisions to ensure its success. I use Zaarly as an example for other software engineering businesses to aspire to become. It demonstrates the importance of communication, prioritization, and trusting one's intuition to achieve success.

Below are a few more critical themes outlined in the Zaarly handbook:

- Output is all that matters, not the number of hours you put in or how much face-time team members have - This means that team members should be focused on producing results and not spending a lot of time in the office.

- The organization should stay as flat as possible - This ensures that people can communicate easily and decisions are made quickly.
- Small teams should form quickly, as needed, and anyone can lead them - This encourages collaboration, flexibility, and innovation by allowing anyone to take the lead when needed.
- Short, clearly addressable projects are preferred - Break big problems down into small projects which can more easily maintain a sense of urgency and momentum over time.
- Employees must have a realistic view of the present but be optimistic about what is possible. It encourages team members to stay grounded and goal-oriented while having a sense of optimism.

These themes emphasize the importance of communication, collaboration, and flexibility for a software engineering culture to be thriving. They are critical components for any software engineering business to strive towards achieving. These five themes are central in helping to ensure success, and Zaarly is an excellent example of how following these principles can lead to positive results.

Zaarly has a structure that is made up of four groups: people who find service providers, people who build the tools to help homeowners and service providers work together, marketers, and people who help those three groups thrive. This structure allows them to reach their goal without becoming overwhelmed by trying to do too much at once. It is summed up perfectly in their famous "money quote," which states:

"Doing one thing extremely well will beat doing five things with mediocrity every day of the week. We've learned that trying and doing everything is the best way to get nothing done."

Looking at how Zaarly has been successful in its business, this approach works. They have focused on a few core areas and done them well rather than attempting to do too much and ending up with mediocre results. This lesson is applied to any software engineering team - focus on getting one thing well instead of trying to accomplish too much. Doing so will help to ensure success and productivity. The five themes outlined in the Zaarly handbook are essential for any software engineering business to strive towards achieving. Utilizing these themes and focusing on a task helps ensure success and productivity for any software engineering team or organization.

8. Valve

Valve, the Steam online game platform creator, is renowned for its revolutionary employee handbook. This document is viral among software engineers and has been widely debated due to Valve's unique vision and culture. The remarkable aspect of this company is its lack of hierarchy or job descriptions. Even Gabe Newell, the founder and president of Valve does not have any direct reports. Every employee can green-light projects and ship products without anyone intervening. This radical approach to the organization allows developers to create the best experiences for customers without any unnecessary barriers getting in their way. At Valve, it is clear that the customer has priority above all else and is essentially the only "boss."

At Valve, team members are given significant autonomy in deciding which projects to consider. As such, project proposals are pitched to the relevant team members to decide if they would like to join and contribute. A proposal must gain more traction and support to be selected and completed. To ensure that this system works, Valve places an immense emphasis on its hiring and firing process. All team members must have a consensus before someone is fired, and even then, they are given a chance to address the issues and turn things around.

This system emphasizes collaboration, trust, and respect for one another to create a thriving software engineering culture. This innovative approach to software engineering truly sets them apart from other companies.

Valve excels in promotions; it uses stack rankings and peer reviews. To ensure fair compensation amongst their team members, they use a system of stack ranking whereby team members are ranked against each other by a set of metrics. In addition, each group conducts peer reviews to provide feedback and guidance on how individuals can improve as individual contributors. It ensures that everyone is held accountable and can identify areas of improvement. This system works in tandem with the lack of hierarchy to support a culture that encourages personal growth and development while offering recognition for hard work. Such an approach has been highly successful for Valve and is something that other software engineering companies can learn from the company.

Themes: Valve prides itself on having a team of responsible, entrepreneurial, and self-managing individuals. They look for these qualities when interviewing potential hires as they believe each person should be capable of running the company. It is evident in their handbook: "Any time you interview a potential hire, you need to ask yourself not only if they're talented or collaborative but also if they're capable of running this company because they will be." This kind of mentality puts a lot of trust and autonomy in its team members, which has greatly benefited their software engineering culture. Other tech companies can learn from Valve's model and replicate it to develop their engineering culture better.

When hiring, Valve looks for what they call a "t-shaped" person. It refers to broad-range generalists with deep expertise in one area. It is an essential factor for them as it ensures team members are flexible and highly knowledgeable about specific business areas. It helps to ensure that everyone can provide high value and specialization in crucial areas of the software engineering process. It is just one example of how Valve has successfully developed its software engineering culture and should be considered by other tech companies looking to do the same.

At Valve, nothing has a permanent structure. Instead, project teams have internal structures tailored to each team's needs at any given moment. It lets team members quickly understand what they are to do when they join a project and provides them with a clear direction. The overall goal is for everyone to constantly look for the most valuable work they can do and then carry it out. Goals for the project, both long-term and short-term, then emerge organically from this philosophy. Valve also encourages their team members to challenge each other's assumptions and decisions - mistakes are acceptable as they see them as learning opportunities. It is through this approach that Valve has been able to build such a thriving software engineering culture. Companies that take similar approaches can benefit in the same way.

Valve recognizes that they are not perfect in every aspect and is clear about the things that could be better. It includes mentoring people, helping those to grow in areas where they are weak, and disseminating information. They even have a quote that sums up their approach: "When you're an entertainment company that's spent the last decade going out of its way to recruit the most intelligent, innovative, talented people on Earth, telling them to sit at a desk and do what they're told obliterates 99 percent of their value." Valve has also created The Valve Handbook for New Employees, which outlines their company culture in detail. This kind of honest self-reflection and commitment to their culture has led to the success of Valve's software engineering teams. Companies that emulate this approach can benefit similarly.

Common themes

When looking at the cultural documents of successful software engineering teams, one central theme emerges: remove all barriers to productivity. It is achieved through practices such as those listed below:

Cutting red tape: Avoiding excessive processes can impede progress.

Hiring responsible people: Recruiting people with sound ethical judgment and time management skills so fewer policies are needed for guidance

Reducing complexity: Utilizing smaller teams, more straightforward solutions, and shorter projects to ensure the team stays calm and can be more adaptable.

Exercise kindness, good communication, and humility: These qualities are integral to productive collaboration and drive data-driven debates.

Allow mistakes: By learning from mistakes instead of dwelling on repercussions, organizations become more innovative and versatile. It allows them to adjust quickly to a changing market.

With the themes of autonomy, mastery, and purpose at the core of each company's employee handbook, some companies have emphasized these values further. Facebook seeks to give its team members a sense of purpose by instilling in them that they are changing the world for the better. Similarly, Etsy fosters an environment where its team members can feel a sense of mastery. It is achieved by allowing new hires to deploy a change on the first day and guiding them through self-improvement experiences throughout their tenure. Lastly, Valve offers team members ultimate autonomy - anyone can work on anything!

Diversity has been and remains a key focus for most companies on this list. Embracing a diverse hiring strategy and workforce provides valuable differences in perspectives and backgrounds, which culminate to make these teams far more effective. The integration of various perspectives to their respective products maximizes creativity, innovation, and overall productivity of the team. By implementing these successful software engineering cultures, companies can foster innovative ideas which will help them stay ahead of the competition.

Chapter 10

Management Style

One of the most significant contributors to a company's success is its management style. The people in charge of overseeing and directing the team members carry significant responsibility. They set the vision, define processes, provide guidance and determine how projects are executed. A company that wants to be successful will need to hire and promote managers capable of performing these tasks at an extremely high level.

Have you ever been in a meeting with a manager whose style made you feel like you were a high school student? If so, it could have been a more productive meeting. Good managers should be able to communicate in ways that ensure their team members understand the vision and goals of the company. They should also be able to share information clearly and adeptly and address problems head-on. These are some key characteristics that make a great software engineering manager.

As we look at how different companies have molded their software engineering managers, we can see many similarities in leadership styles. At the same time, each company has its unique take on what makes a great engineering manager.

Management styles vary significantly across industries, and the best candidates for software engineering managers may differ from those in other industries. A good manager can adapt his style to the needs of his team. For example, the software architecture of a financial services company might require a completely different management style from that of a tech startup. Each manager must speak clearly and effectively with other managers and team members on various topics.

All successful companies have unique management styles that enable them to become world leaders in their industry. Let's look at the different management styles:

1. Coercive Style - "Do what I say"

It is the most autocratic management style and generally results in lower productivity. When a manager employs this style, team members feel disengaged and resentful.

The coercive management style focuses on authority and power, where the manager's role is to dictate to team members what needs to be done without much input from them. The oppressive nature of this kind of management creates a sense of fear among workers and leads to low motivation levels. As a result, team members may put forth minimal effort since they are not given much autonomy in their work and are made to feel as though their opinions do not matter.

In this type of management, the manager will tend to make decisions unilaterally, leading to an environment where team members become passive and unengaged. Furthermore, a coercive leadership style often creates tension between managers and staff members, as team members may feel that they are being micromanaged or maltreated. These further decrease morale within the company, leading to a decrease in productivity levels.

The downside to using this type of management is that it eliminates team members' creativity and innovation since there is little room for input from workers, and ideas are usually suppressed. Therefore, if the tasks set out by the manager are achieved quickly, the lack of team engagement can decrease overall morale and productivity.

For this type of management to be successful, it must be small amounts as it is not conducive to creating an environment where team members feel valued and engaged with their work. The manager should also establish clear expectations and provide support when needed to ensure that the tasks are achieved within a reasonable time frame. Furthermore, the manager must understand that their role is not just about making decisions but also about creating an environment where team members can contribute meaningfully and feel empowered and where there is a sense of ownership over their work.

The coercive management style should be used as something other than the primary form of leadership. Still, it can be beneficial when quick decisions need to be made, and team members need more time to provide input. When using this approach, managers must ensure that they are communicating effectively with their team and providing support where needed. Doing so will help create an environment where everyone feels valued and motivated to contribute meaningfully.

In the rare cases when you need to employ a coercive management style, it's best to make a clear statement that your directive. Simply informing the team that you need a task done a specific way without debate makes it clear that there's little room for creative freedom. At one company, our CTO would often lead the conversation with "I'm putting my CTO hat on..." which meant we needed to disagree and commit to what was being asked of us. In any other case, we knew his ideas were open to discussion or even discarded.

If you've ever had a boss who demanded that you follow her orders without allowing any input from the team, then you know what it's like to work under a coercive manager. While this type of management can be successful in some situations, it is only advisable for use on some occasions since it adversely affects team member's morale and motivation. Instead, managers should focus on creating an environment where team members feel empowered and valued so they can contribute meaningfully to the company's success.

2. Authoritative Style - "Come with me/this way"

If you've ever had a boss who inspired and motivated you to do your best, then you know what it's like to work under an authoritative manager. This type of management seeks to create an environment where team members feel valued and motivated to reach the company's goals.

The authoritative leadership style focuses on collaboration and respect between managers and staff members. It encourages team members to be creative and take the initiative in their work. The manager sets clear expectations for the team and provides support when needed so everyone can contribute meaningfully.

Unlike the coercive management style, the traditional approach focuses on motivation and involvement. When using this type of leadership, managers provide clear direction and encourage team members to take the initiative and contribute their ideas. The manager sets out a vision of how they want things done with guidance and support throughout the process. Allowing team members to be part of the decision-making process makes them feel more engaged and can contribute meaningfully to achieving the company's goals.

The authoritative management style usually leads to higher productivity since team members are given more autonomy and feel valued for their input. This kind of leadership creates an environment where everyone feels empowered to take the initiative and make decisions that will benefit the organization.

Additionally, the team is likely to be more cohesive, and morale will remain high as team members feel that their opinions are being taken into account.

While managers must provide guidance, giving team members enough room to contribute meaningfully is essential. The manager should ensure everyone understands their roles and how they can contribute to achieving the company's goals. Employees should be given opportunities to express themselves without fear of criticism and allowed to take reasonable risks in pursuit of creative solutions.

An authoritative management style is vital when cultivating an engineering culture that motivates and engages its staff members. A manager can create an office environment where everyone feels valued and motivated to contribute meaningfully by providing clear direction while allowing ample room for team member's input. It will lead to higher productivity, morale, and team cohesion—all of which are essential for the success of any engineering team.

3. *Affiliative Style - "People come first"*

When leading engineering teams, an affiliative leadership style often works best. It puts 'people first and is based on creating emotional bonds between the leader and team members through empathy and understanding. The focus here is on building relationships and fostering a sense of unity among the team. It's especially effective for dealing with difficult situations, conflict resolution, and helping team members build trust in each other.

Affiliative Style - "People come first" is a leadership style that helps engineering teams to thrive. It puts people first and is based on creating emotional bonds between the leader and team members through empathy and understanding. This leadership style focuses on developing relationships rather than simply directing tasks or delegating work.

The affiliative style encourages open communication within the team, helping engineers to understand one another's perspectives on projects better while also discovering critical areas for collaboration. Team members are also encouraged to take responsibility for their actions and feel comfortable discussing mistakes or challenges they face. By allowing this kind of open dialogue, the affiliative leader allows their direct reports to learn from their experiences and develop as professionals.

This leadership style is also beneficial for dealing with difficult situations, such as conflict resolution, and helping team members build trust in each other. An effective affiliative leader will ensure that each team member's opinion is heard and respected while guiding the conversation toward a resolution. Through this kind of careful, empathetic mediation, leaders can help bring teams together despite disagreements or tension. Let's look at this practical scenario.

For example, suppose two engineers are working together on a project and have conflicting opinions on how to approach it. An affiliative leader would take the time to discuss each engineer's ideas and then facilitate a conversation between them that allows for open dialogue in a respectful manner. It can help both engineers understand the perspectives of their colleagues while staying focused on finding the best solution for the project as a whole.

In addition, an affiliative leader will work hard to create a sense of unity among their team members by organizing activities like team-building days or simply lunchtime conversations. By creating connections outside work tasks, engineering teams become more closely knit and better equipped to confidently tackle challenging projects.

The affiliative leadership style is based on understanding, trust, and respect. By fostering these positive traits in your team, you can create an environment that encourages growth and collaboration among their engineering teams. This style involves a great deal of patience and effort from the leader, but the rewards are tangible—a happier, more productive team that works together to achieve success.

How can you implement the Affiliative Style as a leader?

Take time to get to know each of your team members and show you care about them as individuals. Ask questions, listen to their responses, and offer support when help is needed. Offer recognition for good work, provide meaningful feedback on progress, and be open to different approaches. Above all else, be patient—building relationships is time-consuming, but it will pay off in the long run.

Encourage open dialogue and collaboration within the team. It means allowing each person to voice their opinion, even if it's different from yours. It also means recognizing when a disagreement arises and working towards resolving it respectfully.

Promote unity amongst your engineers by organizing activities outside of work tasks that bring them closer together and strengthen trust among team members.

Offer support for each engineer as an individual. Please encourage them to take responsibility for their actions, acknowledge mistakes instead of punishing them, and be understanding in difficult situations or when dealing with challenging projects.

Leading engineering teams with an affiliative style can create a more cohesive team dynamic that leads to better outcomes for everyone involved. It's about putting people first and creating trust through understanding and respect. With this approach, leaders can ensure their engineering teams are happy and productive while working together towards common goals.

4. Democratic Style: "Give Employees a Voice."

In engineering, leaders can foster a workplace environment that values collaboration and open communication. Democratic leadership encourages team members to be heard and have a say in important decisions. This type of leadership is based on mutual respect between management and team members, allowing for an open dialogue among team members and managers.

The Benefits of Democratic Leadership in Engineering Teams

Democratic leadership provides multiple benefits to engineering teams by creating an atmosphere for trust, innovation, creativity, and team member's development. When each team member's opinion is valued and respected, it creates an environment where everyone feels that their voice matters. It allows team members to feel comfortable enough to share their ideas and suggestions, leading to new solutions and novel approaches that were previously unconsidered. I like to refer to this as "collective intelligence" because it brings together all team members' collective knowledge and experience.

Additionally, giving team members a say in decision-making processes encourages the creative problem-solving needed to develop solutions to engineering challenges. Democratic leadership also enables team members to grow as professionals through active involvement and management feedback. Engineering teams are better equipped to develop innovative products and processes with this guidance.

Supporting Employee Involvement Through Democratic Leadership

For democratic leadership styles to be effective within an engineering team, management must create an environment that actively supports team member's involvement. It includes providing opportunities for team members to voice their opinions on decisions related to their area of expertise or project goals. Furthermore, it's essential to ensure that team members receive timely feedback on their work and have access to the resources they need to do their job.

It's also vital for leaders to create a workplace atmosphere where everyone feels comfortable speaking up. It means creating an open communication culture and actively encouraging team members to voice their input. Leaders should provide a safe space for honest dialogue, which allows team members to feel confident in sharing ideas without fear of judgment or retribution.

Encouraging Collaboration Through Democratic Leadership

Encouraging collaboration is another critical component of democratic leadership in engineering. Fostering an environment where team members work together towards common goals encourages creative problem-solving and innovative solutions. Leaders must create a culture of trust where team members can openly share ideas and resources.

In turn, this environment allows for more efficient problem-solving, as teams can draw upon each member's strengths and expertise. Moreover, it encourages team members to work together more effectively, resulting in greater efficiency overall. It leads to improved productivity and fewer operational delays or mistakes. You can easily apply this democratic leadership style to various engineering contexts, from small teams working on short-term projects to large-scale initiatives.

When used correctly, democratic leadership can be a powerful tool for creating an effective engineering team. It encourages collaboration, encourages team members to speak up, and supports creative problem-solving. However, leaders must also guide while allowing engineers to take the initiative on their projects. It helps team members stay motivated and develop solutions on their terms, which provides greater job satisfaction and fosters feelings of ownership over their work.

The Pros and Cons of Democratic Leadership

Ultimately, democratic leadership can offer several advantages for engineering teams. Look closely at the pros and cons of this leadership style in engineering teams.

Pros:

- Enables individual contribution from each team member, which leads to more innovative solutions.
- Creates a culture of collaboration and trust.
- Provides team members with an opportunity for professional growth through active involvement in decision-making processes.
- Leads to improved productivity, efficiency, and job satisfaction overall.
- Encourages ownership of their work and greater job satisfaction.

Cons:

- Can be difficult to manage if team members are not aligned on direction or goals.
- Leaders must provide guidance when needed while allowing engineers to take the initiative on their projects.
- Requires careful monitoring by management to ensure that all team members are engaged and participating in the democratic process.
- If not handled correctly, it can lead to resentment among team members who feel their contribution is not valued or rewarded.
- Requires more significant time investment as team members are asked to provide input on decisions.
- May lead to groupthink if left unchecked by management or lack of individual accountability.
- It can be challenging to balance collaboration and individual autonomy.

Democratic decision-making is only effective when there is a clear decision point – a yes or no, a this or that, or a today or tomorrow. You will struggle to agree and resolve the problem if the decision is ambiguous. It is hazardous when the stakes are high, such as choosing a product design direction that may influence the success of your business. In my experience, these ambiguous decisions end up with no clear direction and low motivation, and the outcomes suffer because of it. It's best to save democratic processes for data-driven decisions with clear decision points or reserve them for low-impact issues.

5. Pacesetting Style – "Demand Excellence."

Now, let's take a look at the pacesetting style of leadership. This style is often seen in technological or engineering cultures, where it's essential to stay ahead of the competition and move quickly with innovations. In these settings, leaders are expected to set a high standard for themselves and those they lead - pushing everyone to strive for excellence.

A pacesetting leader will have a relentless drive toward success and take quick action. They may not be as concerned with their team's feelings or ideas but focus more on results, outcomes, and targets. This type of leader sets an ambitious bar that can motivate others to achieve great things but can also cause them to become burnt out from being pushed too hard.

The most successful pacesetting leaders know how to balance high expectations with support, recognition, and appreciation. They must recognize their team's achievements while also demanding that they continue striving for more.

What it looks like in practice?

In an engineering culture, a pacesetting leader will be highly focused on results and timelines. They will take quick action when needed, so the team must be organized and prepared for anything that comes their way. As a leader, you should encourage everyone to work hard and hold themselves and others accountable for results. It would be best to create a safe environment where team members can make mistakes without fear of punishment or criticism. This way, your team will feel comfortable taking risks and trying out new ideas without feeling like they're constantly walking on eggshells.

Be sure to set realistic goals and milestones for yourself and your team. It's important to remember that everyone needs time to learn and grow, so don't expect too much too soon. You also need to reward hard work for people to stay motivated - whether it's a thank you or an extra day off!

The Pros and Cons of Pacesetting Leadership

The pacesetting style has its advantages and disadvantages. I will talk about the pros and cons of being a pacesetting leader.

Pros:

- High standards are set for your team, which can motivate them to reach their highest potential.
- People will be held accountable for results and outcomes.
- Efficiency is improved through quick decision-making and action-taking.
- You will be able to boost morale and motivation through recognition and appreciation.

Cons:

- Your team may become overworked or burnt out if expectations are too high.
- People could feel micromanaged and unappreciated if they're not allowed autonomy.
- Your team may be more productive and motivated with proper recognition and support.
- If you're too focused on results, it could lead to a lack of innovation and creativity.

Pacesetting leadership can be effective if used correctly, but it's important to recognize the potential drawbacks of this style. It's essential to balance high expectations with support and recognition for it to be successful. With the right approach, you can motivate your team to reach their highest potential while creating a healthy and productive working environment.

A great way to look at pacesetting leadership is to see it as a tool in your kit. It's a potent tool that will get results, but it is only usable sometimes due to diminishing returns. In my experience, it's a "crunch time" approach to getting a product out the door or meeting an unusual customer demand that will have a long-lasting positive impact. In short, *crunch time can only be sometime*s, or you will lose people directly via attrition or indirectly via procrastination.

The pacesetting style is just one of many leadership styles, and it's vital to assess what type of leader you are and what type of team you have before committing to a particular style. It is equally vital to remember that every situation is different and may require a different approach. With the right attitude

176

and understanding, pacesetting leadership can be an effective tool for leading your engineering team.

6. Coaching - "Help those who are resistant to change."

You will agree with me that leadership styles within engineering culture can have a range of impacts on the objectives and goals of a team or organization. One leadership style that has become popular in recent years is coaching. The coaching approach to leadership focuses on developing people and using knowledge and encouragement to help those resistant to change.

Coaching can help cultivate an environment of inquiry and creative problem-solving when applied to engineering. This leadership style is particularly effective in managing teams requiring precision and innovation. It has been proven to boost engineers' productivity, morale, and overall job satisfaction.

Understanding the Coaching Style

At its core, coaching is about helping people learn and grow. This leadership style applies to engineering by encouraging engineers to think outside the box and explore new ways of problem-solving. Coaching also promotes an environment of trust and collaboration, allowing engineers to work together towards a common goal.

A vital element of the coaching is enabling team members to take ownership of their development. Instead of a top-down approach, coaching encourages team members to take the initiative and be proactive in achieving their goals. Coaching also promotes open communication between managers and engineers, allowing them to discuss solutions and progress without fear or judgment.

Applying the Coaching Style

When applied to engineering, the coaching style enables managers to lead in a way that encourages team members to learn and grow. For example, instead of giving engineers instructions on how to solve a problem, managers can provide them with guidance and resources. Managers should also give engineers constructive feedback centered around personal growth.

The coaching approach also allows managers to focus on developing relationships with engineers. Managers and engineers should regularly discuss ideas, progress, and challenges. This type of communication allows managers to provide support and resources when needed. For instance, as the leader of a team of engineers working on a complex project, you could tell your team members to bring original ideas and solutions instead of simply giving them orders. You can provide feedback and resources to help them reach their goals more effectively.

Benefits of the Coaching Style

The coaching style of leadership is beneficial for both engineering teams and organizations. As mentioned, it encourages collaboration, innovation, and open communication between managers and engineers. It also enables team members to develop their skills in ways that are patterned to their individual needs.

The Cons of Coaching

Despite the many benefits of coaching, there are some drawbacks. For one, it can take time for managers to provide feedback and support to engineers. Additionally, coaching can lead to disagreements between managers and engineers if there is a need for more understanding and communication. Managers must remember that coaching should be different from traditional instruction and that engineers should still be given clear project expectations.

The coaching style of leadership is a practical approach for engineering teams. It encourages collaboration, innovation, and open communication between managers and engineers. Coaching also enables team members to take ownership of their development, allowing them to learn and grow in ways that cater to their individual needs. With the proper guidance and support, this style of leadership can help teams reach their goals more effectively.

Pro Tip: Have you ever heard of eXtreme Programming, colloquially "pair programming"? It is an excellent tool for a coaching style of leadership because it allows you to work directly with your engineers on solving a problem. Regardless of what you're working on – code, design, architecture – you can use collaborative tools to work in tandem, which allows you to directly coach, remain hands-on and build rapport with your team.

Chapter 11

Teaching Smart People

Our job as leaders and managers is to step down from being the most intelligent person on the team. Instead, we focus on being the most supportive and empowering person. One fundamental way we demonstrate that support is by cultivating a culture of continuous learning. The first challenge is identifying learning opportunities that are interesting to the team and relevant to the business. The stumbling block new leaders face is motivating their team to learn new things and remain current on technology outside their primary responsibilities.

Teaching intelligent people is about recognizing their natural abilities and rewarding them for their strengths. Using this method, you can create a highly trained team of engineers who do the job efficiently.

Teaching Smart People in Engineering

You must understand the differences between teaching and instructing. Teaching actively encourages people to learn, while instructing refers to providing verbal instructions or rules. You should use the teaching approach with engineers struggling with specific topics. In contrast, use instructing for your more intelligent or high-performing engineers who prefer hands-on instructions and feedback. While there is no one-size-fits-all approach to teaching intelligent people, there are some basic strategies to follow. To improve results, incorporate these strategies when assigning tasks and challenges in your teams.

There are two types of intellectuals – those who enjoy learning and are eager and those who are secure in their knowledge (sometimes arrogant) and don't want to devote more energy to pursuing knowledge and experience. Identifying which type your team members fit into and adjusting your strategy for each person is imperative.

Facilitating Learning

To learn effectively, engineers need the right tools and resources. To encourage your engineers to develop their skills, offer them the tools they need to succeed. For example, you can help them by providing access to advanced engineering courses or enrolling them in a local college or university technical program. You can also provide mentors who can give feedback or guidance on their projects and career goals. It would also help if you created an environment where team members feel safe enough to ask questions when unsure about something.

The key to successful learning is letting engineers ask questions and get the tools they need to succeed. Sometimes, sharing knowledge can be challenging because of individual preferences or cultures. For example, some engineers may be more comfortable learning by teaching someone else instead of reading or watching tutorials online. If you encounter this situation, try creating small groups that allow your engineers to share their knowledge with others who want to learn.

Encouraging Learning for All Engineers

It is easy for engineering managers to get caught up in the day-to-day tasks at hand and forget about their teams' development. To stay on top of their professional growth, you should actively educate your engineers and provide them with feedback on their progress. Teaching intelligent people is about recognizing their natural abilities and rewarding them for their strengths. You can combine these strategies to create a team of knowledgeable engineers who do the job efficiently.

Methods of teaching smart people include:

Group learning can be the most effective approach when your team has many engineers because it allows all to participate in discussions and learn from each other. This approach is constructive when engineers want to form a team and attend online technical training together. Team challenges: Team challenges are designed for engineering teams by involving individual challenges requiring teamwork and communication between engineers. These challenges are designed to help engineers learn new skills and get to know each other better. One example is an algorithm-proving challenge requiring engineers to solve the same problem differently.

Technology Guilds are an excellent implementation of group learning. I regularly form guilds on various topics and encourage collaborative learning. For example, I recently set up a certification guild, where a handful of engineers got together to group study for a certification exam. We met bi-weekly to discuss each section of the exam, and one person presented a topic, covering the rote knowledge, a demonstration, and reviewing the relevant sample questions. Every person in the group passed the exam without a struggle after the guild.

Team testing involves engineering managers and engineers working together on joint projects or tasks to measure their results. The results from a team effort can be used to determine the strengths and weaknesses of individual team members and workgroups, as well as test out new technologies or processes. This approach allows engineers to learn from each other and improves efficiency.

One way to implement team testing is by conducting regular hackathons. They can be something other than a full-time distraction, nor happen often. Still, a hackathon is an excellent way for a team to practice objective-driven engineering projects with a conclusion, retrospective, and positive outcome. In addition to the raw educational and cultural value, these are a great way to give closure on a project, e.g., "It's done," which is rare these days.

Team projects: The processes involved in team projects vary significantly according to the needs of each project. However, the most common approaches involve a combination of group brainstorming sessions, individual activities, group-based discussions, presentations, and group testing. Team assignments: Team assignments allow engineers to share ideas and receive feedback on their progress. This approach can benefit teams working on small projects or tasks together.

When I think of team projects, specifically those outside the mainstream of work expected from the team, I think of SkunkWorks Projects. Let's face it – everyone has ideas, but not all are explorable on the company dime. A skunkworks project is an excellent way for a team to work together to mature an idea, perhaps build a proof of concept and prove their idea. If it works, it can easily be turned into a compelling proposal, and if not, it was a learning exercise that didn't have many bureaucratic guardrails.

Teaching by doing: This approach has very similar characteristics to the instructing method described above. In this case, however, engineers learn by doing instead of reading about it online or receiving instructions from managers or coaches. I like to use this approach with my teams because it allows engineers to gain experience by working through problem-solving. Then, they can implement the knowledge they have acquired in similar situations in the future.

Teaching by example: This approach is similar to the "teaching by doing" method described above, but it also involves managers giving feedback to their engineers. Essentially, this method requires engineers to analyze their work and explain how they solved problems. Although this approach may take longer than others, it can be especially effective for more intelligent or high-performing engineers who prefer hands-on instructions and feedback.

Teaching by doing and by example is how I like to bring new team members up to speed or get a junior-level engineer involved in the work. You begin by assigning them simple but relevant tasks – like bugs, test failures, or performance testing. It allows them to learn from exposure and by being hands-on. As they succeed in those tasks, the difficulty and complexity can increase until they're comfortable.

Things to keep in mind when teaching and improving the work of intelligent people include:

Beware of knowledge traps: People of all intelligence levels can fall into knowledge traps. These "traps" occur when team members stop learning because they are afraid to ask questions or admit what they don't know. To prevent this from happening, encourage your engineers to engage with each other and ask questions about topics outside their comfort zones or expertise. You can also help by exposing them to various topics and fields rather than focusing on one area.

Build confidence: Confidence is critical for developing skilled engineers. When team members lack confidence, their skills will be limited by fear of failure or success. A lack of confidence can paralyze engineers, so you should support and mentor your team members. It can improve their confidence and encourage them to voice their concerns in the team setting.

Accountability: Engineering managers can provide accountability by following up on team members' work time, reviewing work assignments, clarifying goals and resources needed, holding individuals or teams accountable for completing tasks or projects, and monitoring team members' performance and job requirements. If a system isn't working as expected, you must communicate this to your team members because they may need to realize the issues themselves.

Learning is not just about solving problems or correcting errors: Learning is a much more significant concept that encompasses challenging assumptions and perspectives. This type of learning leads to understanding, which helps build usable knowledge in the application. This type of learning is developed through cross-functional teams, formal training, or trial and error.

Avoid the "teacher's trap": Teaching a subject that people are familiar with will be much easier than teaching something new. It may make you less motivated and frustrated, especially when you feel like your team members need to understand it. If you're trying to teach your engineers about something new, it's essential to set realistic expectations for their learning ability. Try finding learning materials or resources about the subject on the Internet or create your own by asking an expert.

One strategy I use when avoiding a teacher's trap is delegation. You must keep reinforcing ideas and skills even if much of your team already possesses the knowledge. New people come aboard, take vacations, get sick, and forget or lose context on a given topic. Use this to empower another engineer on your team to take the lead on teaching. It provides them with a new skill: teaching exposes them to more depth of knowledge and gives them confidence in their career development. In addition, the audience gets a fresh perspective and a new voice to the subject – hopefully reinforcing the knowledge.

Feedback is critical: Consistent feedback is essential to the success of your business and your engineering team. Feedback takes many forms: one-on-one discussions, open discussions, code reviews, and retrospectives, to name a few. It's important to keep feedback grounded by the objectives and desired outcomes of the work and avoid nitpicky, personal, or overly critical observations. Be sure to present feedback with the full context: situation, task, action, and result, so you can focus on the impacts of the issue.

Understanding the Impact of Your Behavior: People often overvalue their work, creating a negative feedback cycle. For example, if an engineer does a great job fixing a process error but has yet to receive any positive feedback, he may continue to do the same without improvement. Conversely, poor-quality input with no feedback can create a negative feedback cycle that can affect a team member's perceived self-worth.

Some people are naturally more motivated than their colleagues, and some are naturally more creative. It all comes down to their effort and showing they care about their actions. A great manager will appreciate these behaviors and ensure that these characteristics are encouraged and rewarded when found in team members and co-workers.

It's not about "winning" or "losing": While some people desire to win at all costs, many others do not. By making this distinction, you can learn how your team members respond to setbacks and criticism. Likewise, you will determine how they react when things are going well. This information can help them develop their skills and find the right team.

Being able to give negative feedback: Great managers can provide negative feedback because they know this type of feedback can help their team members learn and grow. On the other hand, bad managers will never provide negative feedback because they fear hurting their team members' feelings or making them feel awful. An effective manager will be able to balance these two aspects of managers: having a "soft" heart, when necessary, with a "tough" mind when communicating with their team members.

Coping with people who avoid learning

Let's face it: Some team members are only motivated to learn about their interests. If you have a team member who is unwilling to learn and grow, there are several things you can do:

Find out the root of their problem – this may be related to their personality and thoughts, feelings, or behaviors. It requires a formal assessment of their motivations, strengths, and weaknesses in leadership and technical skills.

Provide them with some possible solutions to the problem. It should involve seeking out training resources or reading material on appropriate learning activities to help them learn or grow (e.g., cross-functional task forces, formal training, mentorship).

If they're not motivated to learn and grow independently, you must push them. You can push them through their formal performance review or by clearly communicating expectations with your team members.

It is common to see engineers with complacency and security in their current knowledge and skills. An effective leader can influence positive outcomes with continuous learning by applying the proper incentives – tying continuous learning to promotions, bonuses, or extracurricular activities (e.g., conferences).

Do not rush them – forcing someone to "grow up" too fast will cause them to rebel or back down. Instead, you should encourage them slowly and ensure you are always there for support when they need it. When you take this approach, your team members will feel respected and cared for as individuals. They'll also develop a sense of contribution because they know their work matters. It is the best way to lead less motivated people.

Set realistic expectations for performance – especially for people who don't want to learn. You may not get the same results from them, but you will ensure they are challenged to develop their skills and grow as individuals.

Encourage risk-taking: Often, people are afraid of change, whether it's success or failure, and the impact of those results. Encourage a behavior of "failing fast," which helps those who fear failure by having an escape plan. When you encounter someone who is shy or fears the spotlight of success, motivate them by keeping their progress private. In the end, many of us experience the most significant improvements by experiencing failure and working outside our comfort zone, and your teams are no different.

Other things you should know about people who avoid learning

They're scared of the unknown: People who don't want to learn are often fearful of the unknown and are reluctant to explore new things because they don't want to be at a disadvantage or feel awkward or uncomfortable. As a manager, you should remember this when working with these team members: Go slow and remember that your team members are human, just like everyone else. Patience will be necessary for building effective working relationships with these people.

They're susceptible: Most people avoid learning because they feel inferior to their co-workers or boss. As a result, they can become oversensitive and feel threatened by small things. It means you must be careful about how you approach them: Make sure that everything is positive and encouraging when you provide feedback.

They expect perfection: People who are hesitant to learn are often perfectionists. It means that they have difficulty accepting the reality of human error. As a manager, you should remember this when working with these team members: You can't expect them to accept negative feedback without rejecting their ideas or wasting time in training. Instead, you need to be realistic with your expectations and provide positive reinforcement as often as needed for the individual.

They don't want to perform. Another cause for people who are hesitant to grow is that they don't want to perform required tasks. To progress and reach their full potential, they need to learn how things work and take on the responsibility of observing and performing tasks differently rather than avoiding the work altogether. As a manager, you need to provide feedback and guidance so they can feel motivated to grow rather than avoid the work.

They don't want to be vulnerable. People who are hesitant to learn or grow are often afraid of being vulnerable in their working environment. It means they don't want to stand out or attract any attention from others, which may lead to criticism. When it comes down to it, these people want to fit in and not be judged for their mistakes. As a manager, you must ensure your team members know you're supportive of them. They will feel safe enough in this environment and will be more likely to share new ideas or participate in new activities.

They don't want to confront others. This is true in the workplace or personal relationships. As managers, you should try this when working with these types of team members: Offer a small group session or a one-on-one to discuss follow-up questions. While they may not want to ask you, it's better than avoiding the conversation altogether.

They are afraid of being judged. People who don't want to learn or grow tend to be afraid of being judged. To feel okay about their actions, they must realize that they will not affect the performance of others or their superiors. The key lies in making your team members feel comfortable enough to speak up and voice their concerns.

They don't want to stand out. People who are hesitant to learn or grow tend not to want to stand out or make a big deal of themselves, which means they may not be giving their best efforts on the job or projects. When it comes down to it, these people don't want to let anyone down. As a manager, you should remember this when working with these types of team members: Give them positive reinforcement for their efforts. If you don't, they may not make much progress in their goals or tasks at hand.

They're afraid of sharing. This is one of the most recognizable signs that someone is hesitant to learn or grow: They're afraid of sharing their ideas and concerns with others because they may be ridiculed or belittled for doing so. As a manager, you should remember this when working with these types of team members: You must ensure that your team members' ideas are always considered and taken seriously. They need to know that you appreciate their input and that you'll learn from any mistake.

They have low patience in the learning process. Some people are reluctant to learn and grow because they have low patience. They need to realize that growing, learning, and becoming a better person will take time. As a manager, you should remember this when working with these types of team members: You need to make it clear that growth isn't going to happen overnight. They'll have time to perfect their skills and gain new ones as long as they give their best.

Throughout your career as an engineering leader, you will encounter engineers who are ambivalent towards continuous learning. While I have mentioned some strategies thus far for managing these traits in your people, they all share one trait:

you bending over backward for them. The growth of your engineers and yourself need to encourage them to solve these problems independently. Make them aware of the issue, provide options for a resolution and force them to choose a path forward that helps them improve.

Chapter 12

Building High-Performance Teams

You must have heard about the need for high-performance teams in the workplace, and it's true. High-performance teams are groups of people who accomplish a common goal and work efficiently to reach that goal. The key is having team members with complementary skills and a shared vision of success. Here are some tips on how to build and maintain high-performance teams:

Clarifying Your Team's Goal – Research, Development, or Production? – Before building a high-performance team, you must figure out your team's goal. Will your team be responsible for researching and developing new products or ideas? Or will your team be responsible for the production of existing products? Both goals have different expectations and needs and require different people. As a manager, it's crucial to determine the difference between these two teams because they're not going to function similarly.

Clarifying your team's goal will help you create an atmosphere to support that goal. It would help if you were confident that people are working towards the same objective and understand how their role fits into the big picture. It will only be possible if everyone is on the same page.

Everyone must be aware of their strengths. High-performance teams require everyone to understand their unique skills and talents to achieve their goals. Therefore, you need to remind each team member of their strengths so that everyone can work together. I like to use the concept of a "wheel of talent" to help build high-performance teams. The wheel is a list of each member's strengths. For example, if someone performs well at multitasking, that person may be responsible for keeping track of the team's schedule and itineraries.

Everyone must be aware of their weaknesses. It's also crucial for every team member to understand their weaknesses. It will allow them to focus on their strengths and work to improve their weaknesses, if necessary. High-performance teams will only work if each member is comfortable in their skin and understands their unique qualities. I like to use the wheel of talent because it helps me identify people's strengths, weaknesses, and areas of improvement. As a manager, you must organize an annual review to ensure that all team members agree on the company's values.

Everyone must be motivated. High-performance teams require everyone to be motivated for success. It is imperative when you have different working styles or personalities and people tend to hide their weaknesses from each other. An effective way to motivate your team members is by creating a "shared vision" and a shared sense of urgency and responsibility.

Member roles and responsibilities. Once you've determined your team's goal, it's vital to know the roles and responsibilities of every member. It will allow you to discover the weaknesses that need to be addressed or areas for improvement or growth. It is also something that you can analyze during the annual review.

Consider using a simple exercise in creating a RACI matrix – Responsibility, Accountability, Consulted and Informed. It is a matrix that helps you define the roles on your team and who is performing those role functions.

People need to feel connected. High-performance teams usually come together because they want to accomplish something in a group setting. Sometimes, a group can act as "one unit," but sometimes, they don't work that way. It would help if you made sure that everyone understands the "why" for being a team member. The "why" is crucial because it's a driving force for success. As a manager, you should approach your work with high-performance teams by ensuring that everyone knows the team's goals, what they are responsible for, and why they're there in the first place. If there's a person without a role, responsibility, or added value in some way, remove them from the team.

High-Performance Meetings often happen in high-performance teams and other groups and organizations. An excellent way to get everyone to realize their strengths and weaknesses is by using "high-performance meetings."

High-performance meetings need to be clear and concise. Let everyone know their role in the company and how they fit into the picture. Also, you must prove to your team members that you care about their growth and development and encourage them to speak up if something needs to be addressed.

I learned from a great colleague, Jonny, about "Meetings That Don't Suck" during a lightning talk that helped me keep meetings performant. The six rules are: invite the minimum number of people, schedule the least amount of time possible, always have a purpose; set an agenda; define the outcome; and, when the outcome is reached, immediately end the meeting.

Preventing Rivalry: Sometimes, high-performance teams will have rival members to obtain the same position or role. In those cases, it's essential for you as a manager to step in and make sure that these members don't ruin the team dynamic. If your company has a clear policy regarding promotions or goals, try communicating this information with your team members to set realistic expectations for each other.

Disciplining and Encouraging: High-performance teams do require discipline. You also need to relay information so everyone can understand what's happening in the group. Discipline is also essential because it allows the team to work efficiently and reach the common goal with a shared sense of urgency. When a high-performing team becomes successful, you want to reward them by giving them more responsibility and letting them take risks with their projects. As a manager, you need to encourage different team members when they're doing well and encourage others when they have the potential for growth or development.

Driving High Performance: High-performance teams require strong leaders. It would help if you ensured that everyone is motivated and understands the team's goal. An effective way to manage high-performance teams is by ensuring they understand their role and how they're expected to work with each other. Ultimately, you want to be sure that everyone promotes a positive atmosphere through open communication and encouragement.

Like a vision or mission statement, keeping a goal at the top of your mind for your team is a matter of exposure. At important meetings, during presentations, status reviews, retrospectives, demos, etc., always make a point to state the goals. As time progresses, those goals will remain the focus of the engineering efforts, and it will help the team stay grounded in the mission.

Interpersonal skills are essential for a team to work effectively together. This includes being able to discuss ideas constructively, working collaboratively towards a shared goal, providing feedback respectfully, and building trust among the team members. Good interpersonal skills help create an environment where everyone feels comfortable expressing their ideas and working together. It will lead to a higher level of performance overall. For instance, let's say that everyone on your team is working on a project, but they need to communicate with one another. There will be a lack of cooperation, and the project will probably fail. However, if everyone is communicating well with each other, then you'll have better results overall.

Commitment and confidence: High-performance teams are usually made up of highly motivated and committed individuals. If they're not committed, they'll not work hard enough to achieve their goals, and the team will never succeed. Of course, different personalities and skill levels will always exist, but they must commit to the team. Everyone on your team needs to have confidence in one another to commit to the team. They must have confidence in their leadership as well as their colleagues.

Support: High-performance teams rely on support from each other. Everyone needs to understand their role in the team and be able to support other members of the group. If a group member is having problems, then everyone else on the team needs to offer moral support for those members to feel comfortable. If someone is not performing adequately, it's crucial to identify the problem and figure out effective ways to motivate the person to improve. If a team member has personal problems, the rest of the team has to support that person. You can show your support to them by giving them support when they have personal issues or problems at work.

Business results are just one measure of success. Growing a high-performance team should be about growing as individuals instead of focusing on concrete results. It means there should be no place for quantity over quality in high-performance teams; instead, it should be about quality over quantity. It's essential to be sure that the team members know the overall goal of success, but it's also essential for them to take risks and make mistakes. It helps them learn from their mistakes and grow as individuals.

Focusing on individual outcomes may seem counterproductive to your business needs, but it's a superpower. In the book, "The Score Takes Care of Itself," by Bill Walsh, the focus on the individual's needs means the individual can focus on the ball (read: the mission). The learnings from Walsh have helped me prioritize my focus and achieve my business goals with a motivated and healthy team.

Recruiting High-Performance Team Members: When building your team, you must ensure that every member can handle the job. Sometimes, a person might be suitable for the job but may not be the best fit for your team. It would help if you recruited high-performance team members by ensuring they're effective problem solvers and people with good interpersonal skills. It reinforces the idea of hiring for cultural fit – you can train the individual skills like a programming language or tool – as long as you have competent people who get along.

Team building strategies: High-performance teams require trust among the team members. Strong teams are usually made up of people that are comfortable with one another. They feel free to take risks and express their ideas because they know everyone cares for each other.

In today's remote working environment, it's more important than ever to make an effort to support team building. Putting a face to a name and role is a game changer in terms of the creativity and productivity you receive from the team. A regular, in-person team-building activity is necessary to build interpersonal relationships, trust, and alignment that you cannot achieve over a video conference. In my experience, the difference in output after a team has returned from one of these exercises is enormous – you should schedule one today.

Managing conflict: Teams have differences, and when building up high-performance teams, the leader needs to focus on managing conflicts. It includes understanding the causes of conflict (such as how different people work together), using good interpersonal skills by being direct and effective in conflict management, and recognizing when it's time to create change in your team.

Being accountable to the organization: High-performance teams need to know where they stand in meeting their goals, deadlines, and objectives. Teams must be able to create a culture where they push the envelope; however, when they reach the outer limits of their goals, it's then vital for them to push themselves a little bit more, just enough to be effective.

How Big Should Your Team Be? The 80/20 Rule and Logistical Challenges.

Organizing a successful team is essential to achieving success – but the size of that team can be just as important. When considering the team size, it's essential to remember the 80/20 rule: if someone or something isn't needed 80% of the time, then those people should not be part of the team. Although having too many or too few people can both be detrimental to productivity, there are other logistical challenges to consider depending on the size of your team.

Large teams will come with their own unique set of issues. With more people comes a wider variety of opinions and more potential for conflict. Coordinating schedules and availability can also become challenging, especially if certain people are unavailable as often as others. Since there is a greater chance of miscommunication in larger teams, it's essential to have clear lines of communication between everyone involved.

Small teams come with their own set of challenges as well. Unresolved conflicts can become more complex due to the small size of the team, and individual roles may become muddled or unclear. It is also much easier for teams this size to become siloed, with ideas not flowing freely between all members. Therefore, smaller teams need a system for clear communication and idea sharing.

No matter the size of your team, it's essential to be mindful of the potential logistical issues that can arise. Planning and considering how you will approach potential challenges is essential for success. By understanding the 80/20 rule and staying aware of large and small teams' logistical issues, you can ensure that your team is properly organized and functioning.

Use the Definition of Done to Build Alignment, Buy-In, and Foster Ownership.

When teams work together on projects, everyone must have a clear idea of each person responsible for what tasks and how to execute each task. One tool that can help with this process is the "Definition of Done" (DoD). The Definition of Done helps teams to create a shared understanding of who is responsible for which job and when the team can consider their task complete. By creating this definition of what tasks must be completed for a project to move forward, teams can build alignment, buy-in, and foster ownership in the process.

To begin creating a DoD, teams should first identify what needs to be done to complete their project. It might include tasks such as coding elements of software, designing user interfaces, or testing product features. Once the tasks have been identified, each team member should decide who will complete each task and its execution.

When the team has agreed on who is responsible for which job and its execution, they can create a Definition of Done. This definition should outline a task's criteria to be considered complete. The criteria should focus on the quality of work and the timeliness of completion. By making this DoD visible to the whole team, everyone will understand what they are to do to move forward with their project.

Trust and Commitment in a Team Environment - Achieving Mission Goals through Accountability.

In any team environment, accountability is essential to achieving mission goals. While the team leader bears some responsibility for holding members accountable, ultimately, each team member must take ownership of their role in the group. You cannot coerce trust and commitment; instead, they must come from within, and every member must strive to meet expectations with enthusiasm and dedication for the group to succeed.

Creating an atmosphere of mutual respect fosters openness and honesty among members, allowing individuals to hold themselves and one another accountable while working together towards a common goal. For example, if a teammate falls short of meeting deadlines or fails to fulfill assigned tasks, other members should be comfortable voicing their dissatisfaction respectfully and constructively. Doing so will help keep the team focused and on track to meeting its objectives.

The team leader also has a vital role in promoting accountability among members. For instance, setting clear expectations helps ensure that everyone understands their responsibilities and what is expected of them. The leader should monitor progress regularly and maintain open communication with each group member to discuss any issues or concerns. By recognizing successes, fostering collaboration, and providing feedback, when necessary, the leader can help create an atmosphere of mutual respect where accountability flourishes.

Trust and commitment are essential components of a thriving team environment; without them, you cannot achieve accountability. Each member must strive to meet expectations and work together towards a common goal, taking ownership of their role within the group. When this is done effectively, teams can come together to reach even the most ambitious goals.

Chapter 13

Working with Virtual Teams

Globalization has become a primary focus of many businesses, so the need for virtual team members has grown significantly. Some organizations now have virtual members who make up most of their teams.

Virtual teams present unique challenges because there is no physical presence, not even over the phone. It can lead to communication issues, miscommunication, and, ultimately, team failure. The engineering work demands that teams collaborate closely; when some or all teammates are virtual, this can be difficult.

However, the company should employ a few best practices and strategies to ensure that virtual teams are effective and have the highest likelihood of success.

Establish Clear Communication Protocols and Expectations

The cornerstone of any successful virtual team is clear communication. Establishing protocols for regular communication, such as video conferencing or other forms of messaging, will allow members to stay on the same page and ensure that everyone's voice is heard. Additionally, it's important to communicate expectations, particularly regarding deadlines and the quality of work in writing. Working with virtual teams requires greater collaboration between members, which means everyone must have sufficient visibility into the other team members' tasks.

Assign and Communicate with a Project Manager

In virtual teams, assigning a project manager who is a point of contact for team members on technical matters is essential. This person should have experience running similar projects and can act as a go-between for written and verbal communication between teammates.

Often, I see teams operating without any project management. It's an enormous oversight and a disservice to the engineering team responsible for the outcomes of the work. Project managers, even on multiple projects, are essential to the team – they provide productivity boosts by tackling blockers, resolving issues or conflicts that distract the team, and helping with coordination and collaboration that won't happen otherwise.

Ensure Team Members Are Equally Invested in the Project

It's important to remember that working with any virtual team will be challenging. The organization must choose the right team members for each project. Ideally, team members should have the necessary skill sets and resources to complete their projects successfully. But even more importantly, they should be equally invested in helping each other succeed. It means that team members must be willing to assist one another and make the group's success a primary concern.

Involvement outside of the project is also beneficial for establishing a sense of community and camaraderie among team members. It can help cultivate an enjoyable working environment for all involved. If a member isn't feeling it, they should be able to step aside, taking on another job within the company as needed if it will benefit the project as a whole.

Look at Amazon.com, Inc. and its subsidiaries like Amazon Web Services – they allow team members to move between roles quite freely. It keeps team members loyal to the company and gives them opportunities to learn new skills and diversify their knowledge within the business. You can apply this same concept to a team by allowing members to shift their roles. Perhaps your junior software engineer would like to try out DevOps this spring?

Establish a Way of Working

A Ways of Working (WoW) is a document that outlines the parameters of your team's working environment. These documents are established to form a team, and every rule or consideration is made as a group. Refrain from forcing specific issues as a way of working, or it will result in rebellion.

When building out a WoW, consider adding the following information:

- Core working hours - when is the team expected to be online and engaged

- Focus time - when will the team be "heads down" and should not be interrupted

- Vacation / Time Off - how the team will communicate out-of-office events

- Communication - what is the preferred method of communication

- Ceremonies - which ceremonial meetings will occur and when

A successful WoW will be created and ratified as a team. The team must manage themselves, hold each other accountable, and suggest changes to the document. Getting one of these set up is a simple task you can do today - go for it!

Chapter 14

Working with Multicultural Teams

The work environment changes when you consider working with multicultural teams. Multicultural teams consist of people from varying cultures and backgrounds. Different cultures have different communication styles, backgrounds, and emotional make-up, making it challenging to communicate effectively with one another at times. The engineering work demands that teams collaborate closely, and when some or all of the teammates are multicultural, this can be difficult.

However, you can employ a few best practices and strategies to ensure multicultural teams are effective and have the highest likelihood of success.

Understand that people come from different backgrounds with different cultures.

It is an initially crucial step in working with multicultural teams because it's essential to understand the various cultural differences on your team before attempting to work together effectively. These differences may include big-picture issues such as interdependence vs. independence, or they may be more specific concerns like the tendency of western women to desire more direct communication. In contrast, East Asian women tend to prefer indirect communication. Other examples of cultural differences include, but are not limited to:

- The way each culture approaches giving and receiving criticism.
- How each culture feels about leading or following in the workplace.
- The way people are expected to dress, including but not limited to:
  - *Clothing (e.g., revealing vs. modest).*
  - *Grooming (e.g., clean-shaven vs. hairy).*
  - *Accents (e.g., standard American accent vs. thicker British accent).*
  - *Body language (e.g., hands raised vs. hands folded).*
- The preferred foods of each culture (e.g., bacon, burgers, fries, etc.).
- The importance each culture places on the quality of time spent together.
- How people celebrate holidays (e.g., celebrating Christmas vs. celebrating Diwali).
- Different types of humor are used to express different types of sarcasm.
- Different ways people act in public vs. at home and work.
- The level of interest different cultures has in technology and computing.
- The language skills people bring to a team.

At first, communication styles may be difficult to understand because you are working with teammates who may have different communication styles than you, even if they are native English speakers. It is especially true when considering the many communication styles in the English-speaking world.

Understand that communication styles vary between cultures.

It's vital to understand that when working with multicultural teams, the communication style differs depending on who is speaking and their place of origin. These differences may be as simple as the use of facial expressions or body language by westerners versus South Asian people keeping their emotions hidden. It may also be something much more profound, like the western preference for directness in communication being considered rude in Chinese culture, the Japanese love for group consensus, and the Western preference for hierarchical leadership.

Encourage an intercultural attitude among the team.

Intercultural attitudes help to ensure that even if your teammates do not share the same cultural background, they are respectful of each other's culture and willing to learn more about it. This attitude helps bridge cultural gaps in communication and encourages cultural understanding. Leaders must encourage this attitude among their teams because there are many instances when a teammate may make cultural comments about another person's culture, which can be offensive or misunderstood if taken out of context. For instance, you may hear a team member say, "Wow, it's so weird how people in China don't use chopsticks!" From this statement, an intercultural attitude would recognize that this is simply the team member's perception of Chinese culture and not a universal truth.

A leader should encourage an intercultural attitude, especially if they are unaware of their cultural lens as a westerner. If an organization has team members who do not share the same cultural heritage as their teammates, they need to ask questions and learn more than they would have about the various cultures represented on their team.

Encourage cross-cultural training among members of multicultural teams.

Your team members should receive training in cross-cultural communication. The goal of such training is to help your team avoid cultural miscommunication and ensure that when members of a culture try to interact with each other, they are ready for the conversations that come their way. Cross-cultural training can focus on verbal and nonverbal communication but should teach these concepts from various perspectives, including those from other cultures. Understanding the different methods people use to communicate is essential for any successful multicultural team.

Encourage open communication among multicultural teams.

One of the best approaches to a multicultural team is encouraging open dialogue among all members. It means that all team members should be motivated to ask questions and share their viewpoints or ask for clarifications when they need help understanding something.

Mixed nationality teams are usually more motivated, creative, and better at solving problems that often occur in a multicultural team. These teams can be formed by people of different cultural backgrounds working towards the same goal of bringing together the best of each culture. They can also consist of individuals from one or more cultures involved in a project with an international client.

Bi-national teams usually have little communication problems and interpersonal conflicts because they consist of people who are relatively similar to one another. They share similar attitudes and opinions and avoid cultural differences because they speak the same language.

Individuals form bi-cultural teams from different cultures to understand one another more and overcome cultural differences. Individuals also form bi-cultural teams from different backgrounds, nationalities, or all three.

There have been instances where some people have been reluctant to work with others who were not of their national origin because of unfair treatment, racism, and xenophobia. However, nowadays, more people are working together, but managing diversity still needs to be solved. Countries today need to grasp how to manage diversity in the workplace, and managers should consider other cultures, such as religion, race, and age, when hiring new team members.

Time Zones

Globally distributed teams, or even those across the United States, must manage differences in time zones. In addition to multicultural differences, there are nuances to working well with teams that work different hours. Special attention must be made when working in these teams because it is a significant source of conflict. A midday email with a request for urgency may come at an inconvenient time for a teammate. In cultures outside of the west, those who receive these requests might feel obligated to act upon them.

You can manage time zone issues with these simple tricks:

- Put your "core working hours" in your Ways of Working

- Enable time zones in your meeting and communication tools – such as Slack

- Encourage your team members to put their working hours in their calendar

- Discourage your team from working outside of their regular hours

Chapter 15

Dealing With Bias in The Workplace

Now, I want to discuss a crucial topic: dealing with biases in the workplace. It's easy to think of bias as something that only happens outside of the office, but unfortunately, it occurs among colleagues and supervisors.

Before diving into this subject, let's understand what bias is. Bias is "a predisposition to favor one tendency or interpretation in thinking, feeling, or behaving. "So, let me give you an example. It is a person who thinks very highly of herself. She is always confident and comes across as being much surer of herself than even she is. And because of that, she might think that her opinions are more valid than other people's and give her ideas special attention when making decisions.

The first step in dealing with bias in the workplace is understanding what it is and how it can manifest itself. Some common types of bias include ageism, sexism, racism, homophobia, ableism, and religious bigotry. As a manager, it's important to recognize when bias is present in the workplace and take steps to reduce or eliminate its adverse effects.

How does bias set in?

Bias can start with simple prejudices and evolve into more systemic biases. Prejudices are based on first impressions, while systematic bias is rooted in deeply ingrained beliefs that lead to discrimination. Let us look at the types of bias that can occur in the workplace.

Ageism

Ageism can be seen as a prejudice against older team members. Often, managers will see their younger hires as being more motivated, innovative, and having more potential than older workers – usually called "Hot Shots." Age discrimination may also be seen as preferential treatment or discriminatory hiring practices that favor people who are perceived to have more energy, passion, or commitment than people over a specific age. Older workers often face challenges finding new jobs due to age discrimination issues. Even after getting hired, it's common for them to receive lower salaries than their younger counterparts.

Sexism

Sexism is similar to ageism in that it's based on prejudice against a group of people. In the case of sexism, the group in question is made up of women. Sexism can come in the form of preferential treatment or discriminatory hiring practices that favor men over women, as well as condescending remarks and crude jokes while excluding women from meetings or social events. In more extreme cases, it might lead to sexual harassment or assault.

Racism

Racism is bias against people of a different race or ethnicity. It can manifest as offensive jokes, name-calling, slurs, exclusionary practices in hiring or promotion opportunities, or even harassment and targeting based on race. In the workplace, racism can lead to lower morale among team members of color, who may feel that their contributions are not valued by the company or may even be excluded from meetings or social events.

Homophobia

Homophobia is a form of discrimination against people who are gay, lesbian, bisexual, or otherwise have a generally non-heterosexual orientation. People who experience discrimination based on their sexual orientation might be called demeaning names by co-workers denied opportunities for advancement, and have to deal with other forms of emotional abuse.

Ableism

Ableism is discrimination against individuals with disabilities and chronic illnesses. It can include negative attitudes toward people with disabilities, which may lead to discriminatory practices in hiring, promotions, and wages.

Religious Bigotry

Religious bigotry involves discrimination against members of a particular religion or religious group. For example, someone who discriminates against people of a particular faith may call them names or make jokes about their beliefs, suggest that they're unfit for a job because they belong to that faith, or target them in other ways because of their religion. Although this type of behavior is often associated with religious extremists, it can also occur in the workplace and among people who are not affiliated with extremist groups.

Personal bias

It's also important to note that biases rooted in personal preferences and opinions can also become challenging to combat. For example, let's say you're a manager, and all of your team members favor an idea you have, and one person is not. It would help if you immediately gave the idea to the people who support it instead of considering your lone dissenting team member's input.

Effects of bias in the workplace

Management and Leadership decisions are often affected by:

Unconscious bias

One of the significant effects of bias in the workplace is its impact on leadership and management decisions. When people show discriminatory behavior, they don't realize it's happening. Unconscious bias can affect people who don't even consider themselves biased.

As a manager, it's essential to look out for these tendencies and make sure that you aren't making decisions based on personal opinions instead of facts. For example, if one of your team members makes an excellent suggestion but dismisses it because you're not a fan of that person, that could be a sign of unconscious bias.

Bias leads to unfair treatment

Unconscious bias can also lead to unfair treatment of people who may have different views than those of your co-workers or, even worse, different backgrounds than yours. Let's take an example of a manager who doesn't believe that Black people can work in his company. As a result, he may decide to hire only White team members for the project he's about to lead and may even say things that reinforce stereotyping of the Black race. This type of behavior can lead to an imbalance in the workplace and discrimination in hiring practices.

Prejudice leads to loss of talent

Unconscious bias can also lead to the loss of talent because companies will miss out on hiring people who may contribute significantly to their business. The CEO of a large company, known to have a racist view, could be missing out on hiring a competent black woman who holds advanced degrees in Psychology and is the real brains behind her company.

Bias leads to unproductive team members

Several other adverse effects can result from discriminatory workplace behavior. Employees may become frustrated or even hostile when they are put in unfamiliar situations. They may struggle with their work and eventually leave the job, leading to even more losses for the organizations.

Managers must avoid unconscious bias and make fair decisions based on facts instead of personal preferences and biases.

Increased stress levels

Have you ever had a work situation that felt uncomfortable or even unfair? It can be stressful to deal with biases and prejudices in the workplace, primarily when they're not addressed. In many cases, people treated unfairly may become anxious or stressed in their work environment and eventually choose to leave their job because this treatment is just unbearable.

Unconscious bias can also be stressful for those who face discrimination and prejudice at work, making them feel less motivated, anxious, or depressed than their colleagues. For example, a woman working for a company with a history of discrimination towards women in the workplace may become discouraged and eventually consider leaving her job.

Although managers need to be aware of unconscious bias, it's equally important that they make an effort to minimize its effects on their business. For example, if you're a manager at a successful company that discriminates against people based on race or religion, consider choosing a team member training program that will educate your team members and help them develop an awareness of their biases.

Another way you can do this is by encouraging your team members to engage in activities that help them reduce their own bias.

It erodes collaboration

Engineering workplaces encourage collaboration between team members at all levels of the organization. However, when there is a lack of collaboration and respect due to prejudice, it can be difficult for engineers to work together efficiently. For example, an engineer may be offended by a co-worker who always makes jokes based on stereotypes or ethnic biases. This engineer may resent the person and stop working with them as often as he would have in the past.

It makes it difficult to make decisions (Unconscious Stereotypes)

Due to unconscious bias, managers may find it hard to make decisions without considering their personal opinions and biases, which can slow down business growth in the organization. For example, a manager who thinks that all women are incapable of working in his team may continue to hire unqualified female candidates for the positions he has available.

Let's take an example: Let's say all team members in your company are men, and you're about to announce new job openings for new hires. The applicants you receive from your external agency are all females. As a result, you decide to pass on them and hire men instead because you think women need to be more capable of filling the positions.

The problem with this is that by not hiring women to fill the open positions, your company may miss out on hiring candidates who could have significantly contributed to the growth of the business. Instead of basing their hiring decisions on biased opinions, managers need to make fair decisions considering all applicants' qualifications and talents. It will help reveal team members who could contribute significantly to your company's success at all levels of the organization, including women and minorities.

Favoritism in the Workplace

Fraternization and Nepotism - Now, let's consider the case of a manager who prefers people who are like them or share the same religious background. Because of this, they may hire friends and relatives instead of fair and competent candidates who are better suited for the job position.

I see Favoritism in the workplace as a form of unconscious bias. Favoritism is showing partiality toward a favorite team member by offering them special favors or unique opportunities. On the other hand, nepotism refers to hiring or promoting family and friends to positions within an organization.

Favoritism and nepotism can be just as harmful to your business as other types of discrimination, especially if you're an entrepreneur with a small company that only hires one or two team members. For example, in the engineering workplace, favoritism and nepotism can lead to unethical hiring practices and even legal problems for the employer(s) if they're required to pay benefits and other perks to their team members.

Ever heard of an echo chamber? That's what you get when you hire only your friends or relatives. Hiring people they trust in critical positions makes sense with a business leader's livelihood on the line. It's essential to see a pattern of behavior here and stop it from taking over most or all positions. This issue can sneak up on you without warning as your friends hire friends, and before you know it, you've got a reasonably one-dimensional team.

Hoarding Assignments

Another effect of bias in engineering is Hoarding Assignments. It occurs when managers, who prefer one group of team members over another, favor members of their group and pass up opportunities to promote other groups.

For example, suppose you're a manager who prefers engineers with specific educational backgrounds and religious beliefs. In that case, you may stick with your favorites and prevent other engineers from getting the assignments they need to grow in their careers. It can negatively impact your team members, who will miss out on the promotions they need to move up their career ladder.

Abusive behavior

These effects of unconscious bias can also lead to abusive behaviors in the workplace. For example, a manager with an unconscious prejudice against women might engage in behaviors that are seen as harassing or bullying toward female team members. On the other hand, someone who faces bias toward ethnic groups could create a hostile work environment for minority team members.

As a result, team members may feel that they are being discriminated against, and they might feel intimidated or afraid to speak up. Other team members may feel that they're not respected at work, damaging their relationships with colleagues. For example, if you're a manager who is abusive towards women, some female team members may be afraid to approach you or offer suggestions when you need their input for your next project. It could make it challenging to work towards your goals, which could ultimately lead to failure.

Taking into account the effects of bias in the workplace will not only help you identify areas where you need improvement as an individual, but it can also help you improve your company's performance. It will benefit your company because everyone at all levels of the organization can develop their skills and capabilities.

Eliminating bias in the workplace

These are a few things you can do to eliminate bias in your workplace.

Recognize that bias is present

The first step in reducing or eliminating bias is recognizing when it's present and identifying its root source. Sometimes we can unintentionally carry strong biases based on our personal experiences and preferences without realizing it.

Reframe your stereotypes

Instead of thinking about someone based on their gender, ethnicity, sexual orientation, or other traits, try to think about a person as the skills they bring to the table.

Ensure that you're being ethical with your decisions

This can be difficult because we're often forced to make decisions quickly and unconsciously based on biases we've developed over time. When you make a decision: justify it by asking yourself why (and be honest!)

Build a diverse team

Having diversity in your team or hiring a team with people from diverse backgrounds can help you identify the information you might otherwise miss regarding the company's future or development. So many interesting perspectives on usability and product direction come from your team, so it's essential to have a broad view involved in creating it.

Make fair and accurate decisions

This step involves doing an honest job of seeking out all the necessary information available when making a decision. There are many tips on how to do this online.   Furthermore, a general rule of thumb says you should evaluate all applicants reasonably and then justify your decision based on the facts available after doing so.

Apply the "veil of ignorance" when making decisions

You can ensure that your unconscious biases don't influence your decisions. Many organizations are using the "Veil of Ignorance" to weigh essential decisions in the workplace and beyond. This tool is based on a thought experiment by John Rawls in his book titled; A Theory of Justice.

The exercise asks you to imagine that you don't know anything about yourself before being placed into your current situation. It allows you to remain unbiased and make decisions based on available information, not how your biases influence it.

Seek out experts

This may be a more difficult step because it involves asking questions and getting those answers whose role is to make the right decisions based on their experience and knowledge. If you have this information readily available, take the time to review it before making a decision. Leaders fortunate to work for companies with a PeopleOps department should utilize them for this – that is why they are there.

Give people a chance to prove their value

Don't make snap judgments about someone. Always give members of your team the benefit of the doubt by allowing them to prove that they are the right person for the job, regardless of how others perceive them at first glance.

Create a culture of inclusion

Treat everyone around you equally, regardless of race, gender, religion, or sexual orientation.

Reinforce your values

By clearly communicating your standards and expectations to all team members, you can ensure that everyone has the same rules and will feel safe approaching you with any concerns or feedback.

Bias is something that we are all impacted by at a subconscious level. Engineering and computer science fields have vast diversity and represent all types of people, making this a great place to practice being more conscious of our biases and unconscious decision-making. For example, if you decide whether to hire someone or not, you may automatically make a snap judgment based on ethnic background or gender. It can make you biased when hiring someone, which could ultimately cause your company to lose a great new team member. By being aware of the bias in your decision-making, you can actively work towards eliminating it and improving your company's performance.

"Be careful what you tolerate. You are teaching people how to treat you." - unknown.

Dealing with bias in the workplace starts with you. A workplace is not a place to bring about social issues or issues from the outside world. Be sure to nip bias or prejudice issues in the bud immediately and firmly. When people get away with such behavior, it will continue and manifest in unforeseen ways. People leave companies for much less and will undoubtedly leave when disrespected or undervalued.

Epilogue

Leadership and The Business

By Dennis Kelly

Throughout this book, there is a lot of detail built around what leadership is and the intricacies of how it all ties together. However, there is another important part of leadership that has yet to be touched. The part? How leadership affects a business.

I have known Robert Murphy for many years as a peer, leader, and friend. I have worked in various roles throughout the business lifecycle for over 20 years. These roles range from Project Management, Customer Success, Operational Management, and Sales. Most of my experience is based on project outcomes or the effects different things have on the business.

One of my fondest memories of leadership was back when I was in my twenties working as an electrician. Granted, not a lot of people today would see this as a positive, but that's where you need to peel back the different layers and how leaders can mold people. It was summer in Florida, which is always a scorcher, and my job those days was to dig trenches. I was tired, irritated and mad that this is what I was doing. So much so, I quit that day. I walked up to my boss, threw the shovel at him, and said, "this shit is not for me". Then I left. I went home and started to think about what I would do for money, rent, and food. It was a really bad decision. Needless to say, the next morning, I got up and went to work. When I got there, my boss was there and just smiled and asked if I was feeling better. Didn't ask about yesterday, didn't ask why just asked if I was feeling better and ready for a new day. At that moment, he made me feel comfortable that we all have bad days and, most importantly, that real leaders lead by example. That day I learned empathy, that the best way to lead is by trying to relate, compassion and forgiveness.

Let's start this off with one thing that I believe we can agree on: leadership and teamwork among members of a team are the two most essential components to accomplishing any organization's monetary goals. We need to step back and understand that ultimately any business is started to generate some sort of outcome. Most of those outcomes are profitability.

Leadership's biggest effect on the business is the people. If the people are not happy, the business will not thrive. Yes, it might still run, but it will take so much longer for you to reach your goals. When you have inadequate leaders, it ripples throughout everything they touch.

Leadership plays a crucial role in the success of a business and its ability to achieve profitability. Effective leadership can inspire and motivate individuals to work towards a common goal, overcome challenges, and achieve success. In contrast, poor leadership can result in low morale, inefficiency, and, ultimately, decreased profitability.

One of the key ways in which leadership affects the business is through its impact on employee engagement and motivation. When employees feel valued, respected, and supported by their leaders, they are more likely to be engaged and motivated to do their best work. This can lead to higher levels of productivity, which is essential for a successful business. On the other hand, when employees feel unsupported or undervalued, they are more likely to be disengaged and less motivated, which can result in decreased productivity and profitability.

Another way in which leadership affects the business is through its impact on decision-making and problem-solving. Effective leaders are able to make informed and strategic decisions that are in the best interests of the business. They can also inspire and motivate their teams to think creatively and collaboratively, leading to increased innovation and problem-solving. In contrast, poor leaders may make decisions based on personal biases or self-interest, resulting in negative consequences for the business. Poor decision-making and lack of problem-solving can lead to decreased profitability.

Effective leadership also builds and maintains strong relationships with customers, clients, and other stakeholders. Good leaders are able to build trust and rapport through transparent and effective communication, which can lead to increased customer satisfaction and loyalty. This can translate into repeat business and positive word-of-mouth, both of which can contribute to higher profitability. In contrast, poor leadership can result in damaged customer and stakeholder relationships, leading to lost business and decreased profitability.

A study conducted by the Harvard Business Review found that companies with highly effective leaders were four times more likely to have a significant increase in revenue compared to companies with less effective leaders. Similarly, a study by the Boston Consulting Group found that organizations with strong leadership had higher levels of innovation and were more likely to outperform their competitors.

Leadership also affects the business through its impact on company culture. Good leaders are able to create a positive and inclusive work environment that fosters collaboration and innovation. This type of culture can lead to increased job satisfaction and retention of top talent, saving significant business amounts of money and time that would otherwise be spent on recruiting and training new employees. In contrast, poor leadership can create a toxic or negative work environment, leading to high turnover and decreased productivity, ultimately impacting the business's profitability.

Leadership significantly impacts the business and its ability to achieve profitability. Effective leadership can inspire and motivate employees, support strategic decision-making and problem-solving, build strong relationships with customers and stakeholders, and create a positive company culture. These factors all contribute to a successful and profitable business. On the other hand, poor leadership can have negative consequences for the business, leading to decreased productivity, damaged relationships, and a negative work environment, ultimately impacting profitability.

So, when looking at leadership as a whole, you have to think about what is best for you, your team, and the company. It is more challenging than everyone thinks. Being a leader is one of the hardest things people can do. Some people it comes naturally, some they have to work on it and then others just don't have what it takes to lead a team.

When looking at leadership & business together, it's a balance. Sometimes more than the other. The important part is to ensure the leader knows the balance and is able to swing it more from the business side to the people side as needed.

If you take anything away from this chapter, I would like it to be this; when you are leading a team in any capacity, always remember that every decision impacts the business, internally and externally. Take the time to understand how those decisions affect the business overall.

Thanks to my wife Heather and kids Isabella, Dennis, and Emma. Special thanks to Eric Lurie for being my mentor and sounding board.

Conclusion

Leadership with a purpose is about more than just getting people to do what you want them to do. It's about inspiring and motivating your team, developing strong relationships with stakeholders, and creating a positive culture that drives the business forward. Leaders who understand this have the potential to create powerful and profitable organizations that can make a lasting impact on their communities. With the right leadership, the possibilities are endless.

It's time to look at your current leadership strategies and see how you can optimize them for success. Always remember that every decision you make impacts the business and its goals—so be sure to make those decisions with purpose. Good luck!

References

Lessons from 7 highly successful software engineering cultures
Mitch Pronschinske
https://techbeacon.com/app-dev-testing/lessons-7-highly-successful-software-engineering-cultures

Research: How Employee Experience Impacts Your Bottom Line
https://hbr.org/2022/03/research-how-employee-experience-impacts-your-bottom-line

How Diverse Leadership Teams Boost Innovation
Rocío Lorenzo, Nicole Voigt, Miki Tsusaka, Matt Krentz, and Katie Abouzahr. Boston Consulting Group
https://www.bcg.com/publications/2018/how-diverse-leadership-teams-boost-innovation

Coinbase CEO discourages politics at work, offers generous severance to employees who want to quit.
Kate Rooney. CNBC.
https://www.cnbc.com/2020/09/30/coinbase-ceo-offers-severance-to-employees-leaving-over-politics.html

www.ingramcontent.com/pod-product-compliance
Lightning Source LLC
Chambersburg PA
CBHW071648200326
41519CB00012BA/2439